INTERNATIONAL CENTRE FOR MECHANICAL SCIENCES

COURSES AND LECTURES No. 223

THERMOMECHANICS IN SOLIDS

A SYMPOSIUM
HELD AT CISM, UDINE, IN JULY 1974

EDITED BY

W. NOWACKI
UNIVERSITY OF WARSAW

AND

I.N. SNEDDON
UNIVERSITY OF GLASGOW

SPRINGER-VERLAG WIEN GMBH

Originally published by Springer - Verlag Wien New York in 1977

ISBN 978-3-211-81343-0 ISBN 978-3-7091-4354-4 (eBook)
DOI 10.1007/978-3-7091-4354-4

Originally published by Springer - Verlag Wien New York in 1977

ISBN 978-3-211-81343-0 ISBN 978-3-7091-4354-4 (eBook)
DOI 10.1007/978-3-7091-4354-4

INTERNATIONAL CENTRE FOR MECHANICAL SCIENCES

COURSES AND LECTURES No. 223

THERMOMECHANICS IN SOLIDS

A SYMPOSIUM
HELD AT CISM, UDINE, IN JULY 1974

EDITED BY

W. NOWACKI
UNIVERSITY OF WARSAW

AND

I.N. SNEDDON
UNIVERSITY OF GLASGOW

SPRINGER-VERLAG WIEN GMBH

PREFACE

In the past, many fields of Mechanics, including that of Mechanics of Continuous Media, have developed more or less independently of Thermodynamics. At present, however, a stage has been reached where further progress appears impossible without the inclusion of thermodynamic concepts.

In these circumstances, the International Centre for Mechanical Sciences (CISM) decided to present and discuss the results obtained in two course series which were held in 1971 and 1972. In the first one emphasis was placed on the basic concepts and their applications. In particular, the classical approach to thermodynamics of irreversible processes as well as the modern relevant concepts were presented and discussed. In the second series, the developments of the theory of thermal stresses were reviewed and its applications in the field of mechanical, air and spacecraft engineering were surveyed. Besides, attention was focussed on coupled thermoelasticity developed as a synthesis of the theory of elasticity and the theory of thermal conduction. Basic research problems, dispersion of elastic waves, dissipation of energy, etc. were dealt with. Stationary and non-stationary problems, both in the frame of a linear approach and for finite deformations, were analyzed. The lectures were also devoted to thermal disturbances in bodies of various physical response, to thermal effects in piezoelectric media, to the fundamentals of magneto-thermoelasticity and to problems of thermodiffusion in solids.

These course series turned out to be a real success, and also stimulated, creative activities in various research centres. As a consequence, the participants suggested to meet again in about two or three years in order to get acquainted with the new trends of development of thermomechanics.

As a result, a Symposium on "Thermoelasticity" has been held at the CISM in Udine during its Rankine Session, from July 22 to 25, 1974. It has been organized with the aim to survey the steadily growing achievements in this area as well as to discuss questions of further progress.

In order to initiate the discussion, four general reports have been invited. These reports are contained in the present volume. They were scheduled to be read at four consecutive days whereupon the original contributions and papers in the corresponding special fields were delivered. These (fifteen) contributions have subsequently been published in various Journals.

In the first part of this volume Professor I.N. Sneddon presents and discusses the coupled problems of linear thermoelasticity whilst its generalizations and developments in the field of anisotropic elastic media are dealt with, in the second part, by Professor W. Nowacki.

Since in recent years there has been a rapid development in the phenomenological theory of coupled electromagnetic and deformation fields, Professor H. Parkus extends this work to include thermo-magneto-elastic interactions.

Finally, professor C. Woźniak presents a new area of development, the thermoelasticity of nonlinear discrete and continuum constrained systems.

During the Symposium considerable time was devoted to informal discussions, and the Symposium was concluded by a round-table discussion in which the main features of progress in Thermomechanics of Solids have been critically reviewed and treated. The participants agreed on the need of further meetings of a similar kind.

W. Nowacki *W. Olszak.*

Udine, December 1977.

CONTENTS

Page

COUPLED PROBLEMS IN THE LINEAR THEORY OF THERMOELASTICITY
by I.N. Sneddon

THERMAL STRESSES IN ANISOTROPIC BODIES
by W. Nowacki

MAGNETO– AND ELECTROTHERMOELASTICITY
by H. Parkus

ON THE THERMOELASTICITY OF NON-LINEAR DISCRETE AND CONTINUOUS CONSTRAINED SYSTEMS
by C. Wozniak

COUPLED PROBLEMS IN THE LINEAR THEORY OF THERMOELASTICITY

Ian N. Sneddon

University of Glasgow

1 Introduction

The purpose of this introductory lecture is to present a brief account of the linear theory of thermoelasticity starting with a discussion of the basic equations of (non-linear) thermoelasticity and deriving the coupled equations of linear thermoelasticity from them. The treatment leans heavily on the articles [1] and [2].

This is followed by a discussion of the mixed problem of the dynamical theory of linear thermoelasticity and of the variational principles which may be used to derive solutions in special cases.

The next two sections are concerned, respectively, with the propagation of harmonic plane waves in a homogeneous isotropic elastic solid and with a description of some special solutions of the coupled equations.

The survey ends with some remarks on problems with finite wave speed in the heat conduction equation.

2. The basic equations of thermoelasticity

The elastic body B is identified with the bounded regular region of space it occupies in a fixed reference configuration $\mathscr{C}$. The displacement at time t of the point $x \in B$ is denoted by $u(x,t)$. We shall suppose that t belongs to the finite time interval $(0,t_0)$ and we shall write

$$\Omega = B \times (0,t_0), \quad \bar{\Omega} = \bar{B} \times [0,t_0],$$

where $\bar{B}$ denotes the closure of B. By a *motion* of the body we mean a vector field $u \in C^2(\Omega)$ and by the *deformation gradient* the spatial gradient F of the mapping which takes the point x to $x+u(x,t)$, i.e.

(2.1) $$F = 1 + \nabla u \tag{2.1}$$

where 1 denotes the unit tensor and ∇u the gradient of u. It is assumed that the motions under consideration are such that the mapping $x \mapsto x+u(x,t)$ is injective on B and has a smooth inverse so that $\det F \neq 0$.

If we denote the *Piola-Kirchhoff stress tensor* (measured per unit surface area in the reference configuration) by $S(x,t)$ and the *body force* (measured per unit volume in $\mathscr{C}$) by $f(x,t)$ the laws of balance of forces and moments lead to the equations

(2.2) $$\operatorname{div} S + f = 0, \tag{2.2}$$

(2.3) $$SF^T = FS^T, \tag{2.3}$$

where F^T denotes the transpose of the tensor F.

If we denote the *internal energy* (per unit volume in $\mathscr{C}$) by $e(x,t)$, the heat flux vector by $q(x,t)$ and the heat supply, per unit volume in $\mathscr{C}$, by $r(x,t)$ and if we assume that $e \in C^{0,1}(\Omega)$, $q \in C^{1,0}(\Omega)$, $r \in C(\Omega)$ then the local form of the first law of thermodynamics is expressed by the equation

(2.4) $$\dot{e} = S.\dot{F} - \operatorname{div} q + r.$$

Similarly, if $\eta(x,t) \in C^{0,1}(\Omega)$ is the *entropy*, $\theta(x,t) \in C^{1,0}(\Omega)$ is the *absolute temperature* with $\theta(x,t) > 0$, the local form of the second law of thermodynamics is expressed by the inequality

(2.5) $$\dot{\eta} \geq - \operatorname{div}(q/\theta) + r/\theta$$

If we introduce the free energy $\psi = e - \eta\theta$ and the temperature gradient $g = \nabla\theta$ we can write this last inequality in the alternative form

(2.6) $$\dot{\psi} + \eta\dot{\theta} - S.\dot{F} + (g.q)/\theta \leq 0;$$

the inequality (2.6) is called the *local dissipation inequality*..

So far we have made no assumptions concerning the nature of the material forming the body B. Now we assume that the material is *elastic*, that is, that there exist four *constitutive equations* which define ψ, S, η and q as smooth functions of the set of all (F,θ,g,x) for which div $F \neq 0$ and $\theta > 0$. Certain restrictions are imposed on these constitutive equations by the local dissipation inequality, that is, by the second law of thermodynamics. (See, e.g. [2], [4], [5]). It turns out that ψ, S and η are independent of the temperature gradient g and that q satisfies the relation

(2.7) $$(g.q) \leq 0.$$

This last inequality is called *the heat conduction inequality.*

Further, if, for convenience, we omit the variable x and write

(2.8) $$S = \hat{S}(F,\theta), \quad \psi = \hat{\psi}(F,\theta), \quad \eta = \hat{\eta}(F,\theta),$$

we find that $\hat{S}$ and $\hat{\eta}$ can be calculated from $\hat{\psi}$ by means of the

(2.9) $$\hat{S}(F,\theta) = \partial_F\hat{\psi}(F,\theta), \quad \hat{\eta}(F,\theta) = - \partial_\theta\hat{\psi}(F,\theta).$$

The first of these equations is called the *stress relation*, the second is called the *entropy relation* and the equation

(2.10) $$\partial_\theta\hat{S}(F,\theta) = - \partial_F\hat{\eta}(F,\theta)$$

found by eliminating $\hat{\psi}$ between them is called *Maxwell's relation.*

From the definition of the free energy ψ we deduce immediately that the internal energy obeys a constitutive relation $e = \hat{e}(F,\theta)$ where the function $\hat{e}$ is defined by the equation

$$\hat{e}(F,\theta) = \hat{\psi}(F,\theta) + \theta\hat{\eta}(F,\theta). \tag{2.11}$$

Defining the specific heat $c(F,\theta)$ of the material through the equation

$$c(F,\theta) = \partial_\theta \hat{e}(F,\theta) \tag{2.12}$$

we deduce immediately from equations (2.11) and $(2.9)_2$ that

$$c(F,\theta) = \theta\partial_\theta \hat{\eta}(F,\theta) \tag{2.13}$$

We shall confine our attention to materials for which the specific heat is strictly positive, and since, by hypothesis, $\theta > 0$ we deduce immediately from equations (2.12) and (2.13) that the function $\hat{\eta}(F,\theta)$ has a smooth inverse in θ for each choice of F, i.e. that we may write $\theta = \bar{\theta}(F,\eta)$ and hence that we may write the consititutive equations in the alternative forms

$$e = \bar{e}(F,\eta), \quad S = \bar{S}(F,\eta), \qquad \theta = \bar{\theta}(F,\eta), \quad q = \bar{q}(F,\eta) \tag{2.14}$$

The relations (2.9) then imply the pair of equations

$$\bar{S}(F,\eta) = \partial_F \bar{e}(F,\eta), \qquad \bar{\theta}(F,\eta) = \partial_\eta \bar{e}(F,\eta) \tag{2.15}$$

Further conditions on the constitutive equations of an elastic material are obtained by applying the *principle of material frame indifference* (sects. 17-19A of [6]) which states that the constitutive equations are independent of the observer. For this to be so the consititutive equations must have the *reduced forms*

$$\psi = \tilde{\psi}(D,\theta), \quad S = F\tilde{S}(D,\theta), \quad \eta = \tilde{\eta}(D,\theta), \quad q = \tilde{q}(D,\theta,g) \tag{2.16}$$

where

$$D = \frac{1}{2}(F^T F - 1) \tag{2.17}$$

is the *finite strain tensor*, and $\tilde{\psi}$, $\tilde{S}$ and $\tilde{\eta}$ satisfy the equations

$$\tilde{S}(D,\theta) = \partial_D \tilde{\psi}(D,\theta), \qquad \tilde{\eta}(D,\theta) = -\partial_\theta \tilde{\psi}(D,\theta) \tag{2.18}$$

The heat conduction inequality (2.7) also has important consequences. If we define the *conductivity tensor* $K(D,\theta)$ by the equation

(2.19) $$K(D,\theta) = -\partial_g \tilde{q}(D,\theta,0)$$

then as a consequence of (2.7) we have that $K(D,\theta)$ is positive semi-definite, and that

(2.20) $$\tilde{q}(D,\theta,0) = 0\,, \quad \partial_D\tilde{q}(D,\theta,0) = 0\,, \quad \partial_\theta q(D,\theta,0) = 0$$

3. The linear theory of thermoelasticity

We now consider the linear approximation to the system of equations of thermoelasticity consequent upon the following assumptions:-

(a) the displacement gradient ∇u and its time rate of change $\nabla \dot{u}$ are both small;

(b) the temperature field differs only slightly from a prescribed, *uniform* temperature field θ_0, called the *reference temperature;* i.e. $|\vartheta/\theta_0| \ll 1$ where $\vartheta = \theta - \theta_0$;

(c) the time rate of change of the temperature, $\dot{\theta}$, and the temperature gradient g are small.

If $|\nabla u| < \delta_1$, then it is easily that $D = E + O(\delta_1^2)$ as $\delta_1 \to 0$, where E is the *infinitesimal strain tensor* defined by the equation

(3.1) $$E = \frac{1}{2}(\nabla u + \nabla u^T),$$

and similarly if $|\nabla \dot{u}| < \delta_2$, $\dot{D} = \dot{E} + O(\delta_2^2)$ as $\delta_2 \to 0$. Also, if $|\vartheta/\theta_0| < \delta_3$ and $\delta = \max(\delta_1, \delta_3)$ we find that

(3.2) $$\tilde{S}(D,\theta) = C[E] + (\theta - \theta_0)M + O(\delta),\ \delta \to 0$$

where C and M are defined by the equations

(3.3) $$C = \partial_D\tilde{S}(0,\theta_0) = \partial_D^2\tilde{\psi}(0,\theta_0),$$

(3.4) $$M = \partial_\theta\tilde{S}(0,\theta_0) = \partial_\theta\partial_D\tilde{\psi}(0,\theta_0),$$

respectively; the fourth order tensor C is called the *elasticity tensor* and the second order symmetric tensor M is called the *stress temperature tensor.* It should also be noted that, for any pair of symmetric tensors G and H

(3.5) $$G.C[H] = H.C[G];$$

in component form this is equivalent to the symmetry condition

$$C_{k\ell ij} = C_{ijk\ell} \tag{3.6}$$

In a similar way we obtain the approximation

$$q = -Kg \tag{3.7}$$

where

$$K = -\partial_g \tilde{q}(0,\theta_0,0) \tag{3.8}$$

is the *conductivity tensor.* It should be emphasized that there is no reason to believe that, in general, K will be a symmetric tensor; it is always positive semi-definite.

Denoting the density by $\rho(x)$ and the non-inertial body force by b so that $f = b - \rho\ddot{u}$ we see that equation (2.2) becomes

$$\operatorname{div} S + b = \rho\ddot{u} \tag{3.9}$$

Finally, if we introduce the scalar

$$c = \theta_0 \partial_\theta \tilde{\eta}(0,\theta_0) \tag{3.10}$$

– *the specific heat* corresponding to $D = 0$ and $\theta = \theta_0$ - we obtain

$$-\operatorname{div} q + \theta_0 M.\dot{E} + r = c\dot{\theta} \tag{3.11}$$

as the linearized form of the energy equation (2.5).

Collecting these equations together we have :–

The basic equations of the linear theory of thermoelasticity:–

$$\left.\begin{aligned} E &= \frac{1}{2}(\nabla u + \nabla u^T), \\ \operatorname{div} S + b &= \rho\ddot{u}, \\ -\operatorname{div} q + \theta_0 M.\dot{E} + r &= c\dot{\theta} \\ S &= C[E] + (\theta - \theta_0)M \\ q &= -K\nabla\theta \end{aligned}\right\} \tag{3.12}$$

4. The linear theory in the isotropic case

The form of the tensors C,M and K are particularly simple in the isotropic case. (See Sect. 21, 22 and 26 of [1] and [7]). We have

$$\begin{aligned} C[E] &= 2\mu E + \lambda(\mathrm{tr}\,E)1 \\ M &= m1 \qquad (4.1) \\ K &= k1 \end{aligned}$$

where λ and μ are the Lamé constants, k is the conductivity and, in terms of α, the coefficients of thermal expansion

$$m = -(3\lambda + 2\mu)\alpha. \qquad (4.2)$$

We therefore have

The basic equations of linear thermoelasticity for an isotropic body :–

$$\left.\begin{aligned} E &= \frac{1}{2}(\nabla u + \nabla u^T), \\ \mathrm{div}\, S + b &= \rho\ddot{u} \\ -\mathrm{div}\, q + m\Theta_0 \mathrm{tr}\dot{E} + r &= c\dot{\theta}, \\ S &= 2\mu E + \{\lambda(\mathrm{tr}E) + m\vartheta\}1 \\ q &= -k\nabla\theta \end{aligned}\right\} \qquad (4.3)$$

The first two and the last two equations of this system were first derived by Duhamel [8] and later by Neumann [9].
In both cases the strain-rate term

$$m\theta_0 \ \mathrm{tr}E$$

did not appear in the third equation of the set representing the energy balance. There were attempts, at a later date, to justify the inclusion of such a term on the basis of reversible thermodynamics by Voight [10], Jeffreys [11] and Lessen and Duke [12], and on the basis of irreversible thermodynamics by Biot [13]. The derivation outlined here is that given in Sect. 3-8 of [2]; a similar treatment based on modern continuum mechanics and thermodynamics is given in Chap. 8 of Eringen's book [14].

In many applications of the theory two additional assumptions are often made to facilitate the solution of boundary value problems. The first of these – which leads to the *uncoupled theory* – is to assume that in the energy balance equation the term $\theta_0 \operatorname{tr}\dot{E}$ may be neglected so that the temperature field is determined by the pair of equations

$$-\operatorname{div} q + r = c\dot{\theta}, \quad q = -k\nabla\theta.$$

Once the temperature field has been calculated the stress and displacement fields can then be found by use of the first, second and fourth equations of the system (4.3). The second simplifying assumption – which leads to the *quasi-static theory* – is that the inertia term $\rho\ddot{u}$ in the second equation of the system (4.3) may be neglected but that the equations are otherwise unaltered. Indeed, in many engineering applications, in which the geometry is complicated, both approximations are made simultaneously. Such approximate solutions are discussed in the books by Melan and Parkus [15], Boley and Weiner [16], Nowacki [17] and Kovalenko [18].

Here, we shall continue with the discussion of properties of the full set of coupled equations.

If we eliminate S and E from the first, second and fourth of these equations we obtain the equations of motion

$$\mu\Delta u + (\lambda + \mu)\nabla \operatorname{div} u + m\nabla\vartheta + b = \rho\ddot{u}, \tag{4.4}$$

while if we eliminate q and E from the first, third and fifth equations of the set we obtain the coupled heat equation

$$k\Delta\vartheta + m\theta_0 \operatorname{div}\dot{u} + r = c\dot{\vartheta} \tag{4.5}$$

Applying the operator div to both sides of equation (4.4), and the operator curl to both sides of equation (4.5) we obtain the inhomogeneous wave equations

$$\Box_1 \operatorname{div} u = -\rho^{-1}(m\Delta\vartheta + \operatorname{div} b), \tag{4.6}$$

$$\Box_2 \operatorname{curl} u = -\rho^{-1} \operatorname{curl} b, \tag{4.7}$$

in which the operators $\Box_1$, $\Box_2$ are defined by the equations

$$\Box_\alpha f = c_\alpha^2 \Delta f - \ddot{f}, \quad (\alpha = 1,2) \tag{4.8}$$

with

$$c_1^2 = (\lambda + 2\mu)/\rho, \quad c_2^2 = \mu/\rho \tag{4.9}$$

so that c_1 and c_2 are respectively the velocities of the P– and the S– waves in the solid.

Also, if we apply the operator Δ to both sides of equation (4.5) we find that the temperature difference ϑ satisfies the equation

$$kc_1^2\Delta\Delta\vartheta - (m^2\theta_0/\rho + cc_1^2)\Delta\dot{\vartheta} - k\Delta\ddot{\vartheta} + c\dddot{\vartheta} - (m\theta_0/\rho)\mathrm{div}\,\dot{b} + \Box_1 r = 0.$$

Introducing the constants

$$\sigma = cc_1^2 + m^2\theta_0/\rho\ , \quad \kappa_1 = kc_1^2/\sigma\ , \quad \kappa_2 = k/c, \tag{4.10}$$

we see that we can write this equation in the form

$$\sigma\kappa_1\Delta\Delta\vartheta - \sigma\Delta\dot{\vartheta} - k\Delta\ddot{\vartheta} + c\dddot{\vartheta} = (m\theta_0/\rho)\mathrm{div}\,\dot{b} - \Box_1 r \tag{4.11}$$

In the case in which there are no non-inertial body-forces and no heat-sources this reduces to

$$\sigma\kappa_1\Delta\Delta\vartheta - \sigma\Delta\dot{\vartheta} - k\Delta\ddot{\vartheta} + c\dddot{\vartheta} = 0. \tag{4.12}$$

If we now introduce the diffusion operators

$$\mathscr{D}_\alpha = \kappa_\alpha\Delta - \partial/\partial t, \qquad (\alpha = 1,2) \tag{4.13}$$

We see that equation (4.13) can be written either in the form

$$\left(\sigma\Delta\mathscr{D}_1 - c\frac{\partial^2}{\partial t^2}\mathscr{D}_2\right)\vartheta = 0 \tag{4.14}$$

or in the form

$$\left(c\mathscr{D}_2\Box_1 - (m^2\theta_0/\rho)\frac{\partial}{\partial t}\Delta\right)\vartheta = 0. \tag{4.15}$$

In a similar way we can show that the displacement field u(x,t) satisfies the equation

$$\Box_2\left(\sigma\Delta\mathscr{D}_1 - c\frac{\partial^2}{\partial^{t2}}\mathscr{D}_2\right)u = 0 \tag{4.16}$$

again in the case $b = 0$, $r = 0$. (Cf. [19]). This is the analogue in the linear theory of therme ïasticity of the well-known equation

$$\Box_1\Box_2 u = 0$$

in the isothermal case.

The field equations may be uncoupled in a different way based on the Helmholtz resolution

$$b/\rho = -\nabla\chi - \text{curl}\,\boldsymbol{\gamma} \tag{4.17}$$

of the body force, where if $\mathbf{b} \in C^{2,1}(\Omega) \cap (\bar{\Omega})$, $\dot{\chi}$ and $\boldsymbol{\gamma}$ both belong to $C^{2,1}(\Omega)$. If we make a similar decomposition

$$\mathbf{u} = \nabla\phi + \text{curl}\,\Psi \tag{4.18}$$

of the displacement vector, we find that we can satisfy the equation of motion by choosing ϕ and ψ to be solutions of the inhomogeneous wave equations

$$\Box_1\phi = \chi - m\vartheta/\rho, \quad \Box_2\Psi = \boldsymbol{\gamma}$$

If we solve the first of these for ϑ we obtain

$$\vartheta = (\rho/m)(\chi - \Box_1\phi) \tag{4.19}$$

and substituting from this equation and equation (4.18) into equation (4.5) we find that

$$c\mathscr{D}_2\Box_1\phi - (m^2\theta_0/\rho)\Delta\dot{\phi} = c\mathscr{D}_2\chi + (mr/\rho) \tag{4.20}$$

Hence equations (4.18) and (4.19) yield a solution of the coupled equations of thermoelasticity in the isotropic case if ϕ is a solution of (4.20) and ψ is a solution of the inhomogeneous wave equation

$$\Box_2\Psi = \boldsymbol{\gamma}\,. \tag{4.21}$$

This solution is known as the *Deresiewicz-Zorski solution*; c.f. [20] and [21]. The completeness of this solution has been established by Sternberg [22]; see also [23].

An analogue of the Cauchy-Kovaleski-Somigliana solution of the isothermal theory (cf. (67.4) of [1]) has been discussed by Podstrigach [24], Nowacki [25] and Rüdiger [26] but, because of its rather limited use, we shall not consider it further here.

5. The mixed problem of the dynamic theory of thermoelasticity

In this section we look briefly at a certain class of boundary-initial-value problem for the equations of linear thermoelasticity. We assume that the boundary ∂B can be decomposed into four non-intersecting subsurfaces $\mathscr{S}_j$ ($j = 1, \ldots 4$), i.e.

that

$$\partial B = \bigcup_{j=1}^{4} \mathscr{S}_j \, , \, \mathscr{S}_j \cap \mathscr{S}_k = \mathscr{S}_j \delta_{jk}$$

and we denote $\mathscr{S}_j \times [0, t_0]$ by $\mathscr{I}_j$.

The *mixed problem* of the dynamic theory of thermoelasticity is that of finding a set* $[u, E, S\ \vartheta, g\ , q]$ corresponding to the body force b and the heat supply r such that the equations (3.12) are satisfied and, in addition

(5.1) $u(x,0) = u_0(x), \quad \dot{u}(x,0) = v_0(x), \quad \vartheta(x,0) = \vartheta_0(x), \ \forall\, x \in \bar{B};$

(5.2) $u(x,t) = \hat{u}(x,t), \ \forall (x,t) \in \mathscr{I}_1;$

(5.3) $s(x,t) \equiv (Sn)(x,t) = \hat{s}(x,t), \ \forall (x,t) \in \mathscr{I}_2;$

(5.4) $\vartheta(x,t) = \hat{\vartheta}(x,t), \ \forall (x,t) \in \mathscr{I}_3;$

(5.5) $q(x,t) \equiv (q.n)(x,t) = \hat{q}(x,t), \ \forall (x,t) \in \mathscr{I}_4.$

In these equations the initial displacement u_0 , velocity v_0 and temperature ϑ_0 , the surface displacement $\hat{u}$, traction $\hat{s}$, temperature $\hat{\theta}$ and heat flux $\hat{q}$ are prescribed. If such a thermoelastic process exists, it is called a *solution of the mixed problem.*

It has been shown by Ionescu-Cazimir that the solution of the mixed problem is unique if the elasticity tensor C is positive semi-definite and the specific heat c is strictly positive [27] ; the proof of the uniqueness theorem in the case of an isotropic solid has been given previously by Weiner [28]. In the proof of uniqueness an important part is played by the total energy

(5.6) $$\mathscr{U}(t) = \frac{1}{2}\int_B (\rho \dot{u}^2 + E.C[E] + c\vartheta^2/\theta_0)\,dv$$

(*) Such a set, with additional assumptions concerning the differentiability properties of the components, is called a *thermoelastic process corresponding to* [b, s] *and* [r, q].

and by the fact that if $[u,E,S,\vartheta,g,q]$ is a thermodynamic process corresponding to $[0,s]$ and $[0,q]$ with $s.\dot{u} = q\vartheta = 0$ on $\partial B \times [0,t_0)$, then

$$\mathcal{U}(t) \leq U(0), \quad 0 \leq t < t_0 \tag{5.7}$$

Knops and Payne [29] have proved the uniqueness of the solution without the hypothesis that C is positive semi-definite and have also shown that the solution depends continously on the initial date. That C need not be positive semi-definite was shown also by Brun [30]. Existence, uniqueness and asymptotic stability theorems relevant to the mixed problem have been derived by Dafermos [31].

6. Variational principles

To derive variational principles in linear thermoelasticity we introduce the concept of *an admissible process.* By an admissible process we mean an ordered array

$$p = [u,E,S,\vartheta,g,q]$$

where, in the usual sense of the theory of elasticity, u is an admissible motion and S is a symmetric tensor with certain continuity and differentiability properties, and similarly for E ; ϑ is a continuously differentiable scalar field, g is a continuous vector field and q is a continuously differentiable vector field (with domain $\bar{\Omega}$ in each case). The fields forming the components of p need not be related.

The variational principles concern functionals $\Lambda : \mathscr{A} \to \mathbb{R}$, where $\mathscr{A}$ is the space of all admissible processes. We define

$$\delta_{p^*} \Lambda\{p\} = \frac{d}{d\lambda} \Lambda\,(p + \lambda p^*)\Big|_{\lambda=0}$$

where $p + \lambda p^* \in \mathscr{A}$ for every scalar λ. If

$$\delta_{p^*} \Lambda\,\{p\} = 0,$$

for every p^* meeting the above requirement we write

$$\delta\Lambda\,\{p\} = 0.$$

In the definitions of the relevant functionals, it is convenient to use the notation

$$B(x,t) = i*b(x,t) + \rho(x)[u_0(x) + tv_0(x)]\,,$$

$$R(x,t) = 1*r(x,t) + c(x)\vartheta_0(x) - \theta_0 M(x).E_0(x)\,,$$

where $\mathbf{u}_0$, $\mathbf{v}_0$ and ϑ_0 have the meanings previously assigned to them, $E_0(x) = E(x,0)$ and $f*g\,(x,t)$ denotes the convolution

$$\int_0^t f(x,t-\tau)\,g(x,\tau)\,d\tau \ ;$$

the functions i and 1 are defined by

$$1(t) \equiv 1, \quad i(t) \equiv t, \quad \forall t.$$

We define the functionals Λ_t , θ_t, Φ_t on $\mathscr{A}$ the set of all admissible processes and for $t \in [0,t_0)$ by the equations

$$\Lambda_t\{p\} = \int_B \left[\frac{1}{2}\, i*E*C[E] + \frac{1}{2}\rho u*u - i*S*E - (i*\mathrm{div}S + B)*u\right]dv$$
$$- \frac{1}{\theta_0}\int_B i*\left[\frac{1}{2}g*Kg + \frac{1}{2}c\vartheta*\vartheta + 1*q*g + (1*\mathrm{div}q - \theta_0 M.E + R)*\vartheta\right]dv$$
$$+ \int_{\mathscr{S}_1} i*s*\hat{u}\,da + \int_{\mathscr{S}_2} i*(s-\hat{s})*u\,da + \frac{1}{\theta_0}\int_{\mathscr{S}_3} i*1*q*\hat{\vartheta}\,da$$
$$+ \frac{1}{\theta_0}\int_{\mathscr{S}_4} i*1*(q-\hat{q})*\vartheta\,da \ ;$$

where A and c' are defined by the equations

$$A = -C^{-1}[M] \ , \qquad c' = c - \theta_0 M.A \ ;$$

$$\Phi_t\{p\} = \int_B \left(\frac{1}{2}\, i*S*E + \frac{1}{2}\rho u*u - B*u\right)dv$$
$$+ \frac{1}{\theta_0}\int_B i*\left(\frac{1}{2}g*q - \frac{1}{2}c\vartheta*\vartheta * \frac{1}{2}\theta_0 (M.E)*\vartheta - R*\vartheta\right)dv$$
$$- \int_{\mathscr{S}_2} i*\hat{s}*u\,da - \frac{1}{\theta_0}\int_{\mathscr{S}_4} i*1*\hat{q}*\vartheta\,da.$$

In the first variational principle (see [32] [33] and [34] and Sect.24 of [2]), it is not required that the admissible processes satisfy any of the field equations, or any of the initial and boundary conditions. We have:–

First Variational Principle: $\delta\Lambda_t\{p\} = 0, (0 \leqslant t < t_0)$ *at an admissible process* $p \in \mathscr{A}$ *if and only if* p *is a solution of the mixed problem.*

On the other hand, in the second variational principle restrictions are imposed on the domain of the function θ_t. We have :–

Second Variational Principle: Suppose that the elasticity tensor C *and the conductivity tensor* K *are both invertible and that* $\mathcal{A}$ *is the set of all admissible processes that satisfy the strain-displacement and the thermal gradient-temperature relations. Then* $\delta\theta_t\{p\} = 0$ *, at* $p \in \mathcal{A}, 0 \leq t < t_0$ *if and only if* p *is a solution of the mixed problem.*

To formulate the third variational principle we place further restrictions on the processes p ; by a *kinematically and thermally admissible process* we mean an admissible process which satisfies

(i) the strain-displacement relation,
(ii) the thermal gradient-temperature relation,
(iii) the stress-strain-temperature equation,
(iv) the heat conduction equation,
(v) the displacement and temperature boundary conditions.

In terms of this definition, we have the third variational principle:–

The Principle of Minimum Potential Energy: If $\mathcal{A}$ *denotes the set of all kinematically and thermally admissible processes, then* $\delta\Phi_t\{p\} = 0$ *at* $p \in \mathcal{A}, 0 \leq t < t_0$ *if and only if* p *is a solution of the mixed problem.*

7. Thermodynamic waves

A considerable amount of work has been done on the propagation of harmonic plane progressive waves in a homogeneous isotropic elastic solid with neither heat sources nor body forces acting within it – see, for example, [35] – [38], [43] and Chapt. 2 of [39].

If a and p are unit vectors and Λ, θ, γ and ω are constants

$$u = U a \exp[i(\gamma p.x - \omega t)], \quad \vartheta = \Theta \exp[i(\gamma p.x - \omega t)] \qquad (7.1)$$

represent plane progressive ways propagating in the direction p; the direction a is called the *direction of displacement.* Substituting from equations (7.1) into equations (4.4) and (4.5), but with $b = 0$ and $r = 0$, we find that the constants must satisfy the equations

$$(7.2)\qquad \left.\begin{aligned}(\rho\omega^2-\mu\gamma^2)U\mathbf{a} - [(\lambda+\mu)(\mathbf{a}.\mathbf{p})\gamma^2 U - im\gamma\Theta]\mathbf{p} &= 0\\ m\theta_0\gamma\omega(\mathbf{a}.\mathbf{p})U + (ic\omega - k\gamma^2)\Theta &= 0\end{aligned}\right\}$$

If $(\mathbf{a}.\mathbf{p}) = 0$ we get the *transverse waves* and it follows from this last pair of equations that $\Theta = 0$ and $\rho\omega^2 = \mu\gamma^2$. Since the speed of propagation of the waves (7.1) is $\mathrm{Re}(\omega/\gamma)$ we conclude that *the transverse waves are independent of thermal effects and propagate with speed* $\sqrt{(\mu/\rho)}$.

On the other hand if $\mathbf{a} = \mathbf{p}$ we get the *longitudinal waves* and it follows from the equations (7.2) that for the longitudinal waves U and Θ satisfy the equations

$$(\rho\omega^2 - \mu\gamma^2)U - (\lambda+\mu)\gamma^2 U + im\gamma\Theta = 0$$

$$m\theta_0\gamma\omega U + (ic\omega - k\gamma^2)\Theta = 0.$$

Eliminating the ratio U/Θ from these equations we get the equation

$$(7.3)\qquad (\omega^2 - c_1^2\gamma^2)(\omega + i\kappa_2\gamma) - \Gamma\omega\gamma^2 = 0$$

connecting ω and γ. In this equation c_1 and k_2 are defined by equations (4.9) and (4.6) respectively and Γ is defined by the equation

$$(7.4)\qquad \Gamma = (m^2\rho\theta_0/c)$$

If we assume that ω is given (and real), we can solve equation (7.3) for γ; we find γ is complex so that the longitudinal waves are attenuated and, further, since the speed of propagation is a function of ω, the waves are *dispersed*.

On the other hand, if we assume that γ is given (and real), we can solve equation (7.3) for ω; we find that ω is then complex so that the waves are damped and dispersed.

The appropriate numerical work is presented, and discussed, in [43] and Chapt.2 of [39]; the latter reference also contains an account of the work of Lockett on the propagation of *thermoelastic Rayleigh waves* along the plane boundary of a half-space.

8. Some special solutions of the coupled equations

When the difficulty of solving initial-value, boundary-value problems in classical elastodynamics is recalled, it is hardly surprising to find that the derivation of closed form solutions of even the simplest kind of mixed problem (in the sense of §5) is very difficult. In this section we shall briefly refer to some attempts to derive such solutions.

The determination of the displacement $u(x,t)$, the stress $\sigma(x,t)$ and the temperature $\vartheta(x,t)$ in a semi-infinite rod $x > 0$ is considered in [40] when the end $x = 0$ is disturbed by the application of stress or temperature. The boundary conditions are of the six basic types

$$\begin{aligned}
&\text{(i)} & \vartheta(0,t) &= \vartheta_0(t), & u(0,t) &= 0;\\
&\text{(ii)} & \vartheta(0,t) &= \vartheta_0(t), & \sigma(0,t) &= 0;\\
&\text{(iii)} & \sigma(0,t) &= \sigma_0(t), & q(0,t) &= 0;\\
&\text{(iv)} & \sigma(0,t) &= \sigma_0(t), & \vartheta(0,t) &= 0;\\
&\text{(v)} & u(0,t) &= u_0(t), & q(0,t) &= 0;\\
&\text{(vi)} & u(0,t) &= u_0(t), & \vartheta(0,t) &= 0.
\end{aligned}$$

In these equations

$$q(x,t) = \frac{\partial\vartheta(x,t)}{\partial x}$$

and the functions ϑ_0, σ_0 and u_0 are prescribed on the positive real line. The solutions of these problems were derived by means of a systematic use of Fourier and Laplace transforms.

Similar methods are used to determine the state of stress and the distribution of temperature in the *finite* rod $0 < x < \ell$ when the end $x = \ell$ is subjected to the boundary condition

$$\sigma(\ell,t) = 0, \quad q(\ell,t) = 0, \quad t > 0.$$

The problem of determining the stresses produced in an infinite elastic solide by uneven heating is discussed in [41]. Here the problem is to solve the coupled equations (4.3) for the region $\mathbb{R}^3 \times (0,\infty)$ when $b \equiv 0$ and the function r is prescribed. The general solution is found by the method of Fourier transforms, and the solutions for plane strain and for the quasi-static case are deduced from it. The solution of the corresponding axisymmetric problem is found by using a multiple integral transform which is a Hankel transform over the cylindrical coordinate ρ and a double Fourier transform over the variables z and t. The following special cases are considered:

(i) the stress due to a periodic line source;

(ii) the effect of a line source moving with uniform velocity in a direction perpendicular to its own length;

(iii) the stress due to an impulsive line source,

(iv) the stress due to an impulsive point source.

The corresponding problems for a half-space were also considered.

Lockett and Sneddon [42] discussed the complementary problem of determining the generation of thermoelastic disturbances in an infinite elastic solid by body forces, i.e. of solving the coupled equations (4.3) in $\mathbb{R}^3 \times [0,\infty)$ when b is prescribed and $r \equiv 0$. Again the method of integral transforms is used.

The problem of determining the distribution of stress induced in an infinite elastic solid with a spherical cavity, when the face of the cavity is subjected to a thermal as well as a mechanical constraint, was considered by Chadwick [43]. In problems of this type the calculation of the inverse Laplace transforms is very complicated and recourse has often to be made to approximate methods of solution. Two such methods are: *(a) perturbation expansion* and *(b) asymtotic expansion for small times.*

9. Problems with finite wave speed in the heat conduction equation

So far we have assumed that the constitutive equation for q is the *Fourier Law*

$$q = -K\nabla\vartheta \tag{9.1}$$

which when substituted into the third equation of the set (3.12) leads to an equation involving the temperature and the strain. For instance, in the isotropic case we obtain the equation (4.5) which, by the elimination of the strain terms leads to the temperature equation (4.7). An even simpler situation is presented in the *uncoupled theory* where, in the absence of heat sources, the third equation of the set (3.12) reduces to

$$-\operatorname{div} q = c\dot{\vartheta} \tag{9.2}$$

so that ϑ satisfies the heat conduction equation

$$\operatorname{div}(K\nabla\vartheta) = c\dot{\vartheta}$$

or, in the isotropic case

$$\operatorname{div}(k\nabla\vartheta) = c\dot{\vartheta}$$

or

$$k\Delta\vartheta + \nabla k.\nabla\vartheta = c\dot{\vartheta} \tag{9.3}$$

which is a parabolic partial differential equation with infinite signal speed.

Since the idea of a thermal disturbance being propagated with infinite speed is unacceptable, there have recently been attempts to modify Fourier's law (9.1) to provide for a finite signal time. See, for instance, the papers by Meixner [45], Müller [46], Gurtin and Pipkin [47]. A very simple model by Roetman [48] is also worthy of study.

Suppose for simplicity that equation (9.1) is replaced by an equation of the form

$$q + \tau_0\dot{q} = - k\nabla\vartheta \tag{9.4}$$

(in the isotropic case) where τ_0 is a constant with the dimensions of time. Then operating on the both sides of this equation with div and making use of equation (9.2) we find that ϑ now satisfies the hyperbolic equation

$$\left[\Delta - \frac{1}{v^2}\frac{\partial^2}{\partial t^2}\right]\vartheta = \frac{1}{\kappa_2}\frac{\partial\vartheta}{\partial t} + k^{-1}(\nabla k.\nabla\vartheta) \tag{9.5}$$

where k_2 is defined by the third equation of the set (4.6) and the wave velocity v is defined by the equation

$$v^2 = k/c\tau_0 \tag{9.6}$$

In this way we see that the inclusion of the term $\tau_0 q$ in equation (9.4), if it can be justified on physical grounds, will provide for a finite signal time for heat propagation. It would appear that the Fourier law should be generalized to some such form as

$$q + \tau_0\dot{q} + Qq = - K\nabla\vartheta \tag{9.7}$$

Where Q and K are second order tensors which are characteristic of the material forming the body B. When equation (9.7) replaces the final equation of the set (3.12) the resulting system of equations will be even more difficult to solve.

Finally, it is of interest to note that Fulks and Guenther [49] have demonstrated that in the singular perturbation problem in which v^{-2} in equation (9.5) is small, the solution of (9.5) agrees very closely with that of equation (9.3) except when a heat pulse is introduced. A front then propagates outwards from the pulse until after passing a given point the solution of (9.5) begins rapidly to settle down to that of (9.3). It is precisely during the passage of that front that we should expect the elastic material to respond dynamically to the change in temeperature.

REFERENCES

[1] GURTIN,M.E., The Linear Theory of Elasticity: In Handbuch der Physik, edit. by C. Truesdell (Springer, Berlin, 1972).

[2] CARLSON, D.E., Linear Thermoelasticity: In Handbuch der Physik, edit. by C. Truesdell (Springer, Berlin, 1972).

[3] COLEMAN, B.D., & NOLL, W., Arch. Rational Mech. Anal. 13 (1963), 167.

[4] COLEMAN, B.D., & MIZEL, V.J., J. Chem. Phys. 40 (1964), 1116.

[5] GURTIN, M.E., Arch. Rational Mech. Anal. 19 (1965), 339.

[6] TRUESDELL, C., & NOLL, W., The Non-Linear Field Theories of Mechanics. In Handbuch der Physik, III/3, edit. by S. Flügge (Springer, Berlin, 1965).

[7] COLEMAN, B.D., & MIZEL, V.J., Arch. Rational Mech. Anal. 13 (1963), 245.

[8] DUHAMEL, J.M.C., J. Ecole Polytechn. 15(1837) 1.

[9] NEUMANN, F.E., Abhandl. k. Akad. Wiss. zu Berlin (1841) 2. Teil,1.
see also: NEUMANN, F.E., Vorlesungen über die Theorie der Elasticität, (Teubner, Leipzig, 1885).

[10] VOIGT, W., Kompendium der theoretischen Physik, Bd. 1 (von Veit, Leipzig, 1895).

[11] JEFFREYS, H., Proc. Cambridge Phil. Soc. 26 (1930), 101.

[12] LESSEN, M. & DUKE, C.E., Proc. 1st Midwest Conf. on Solid Mech. pp. 14-18; (University of Illinois, 1953).

[13] BIOT, M.A., J.Appl. Phys. 27(1956), 240.

[14] ERINGEN, A.C., Mechanics of Continua, (Wiley, New York, 1967).

[15] MELAN, E. & PARKUS, H., Wärmespannungen, (Springer, Wien, 1953).

[16] BOLEY, B.A., & WEINER, H.J., Theory of Thermal Stresses, (Wiley, New York, 1960).

[17] NOWACKI, W., Thermoelasticity, (Pergamon, Oxford, 1962).

[18] KOVALENKO, A.D., Thermoelasticity, (Wolters-Noordhoof, Gröningen, 1969).

[19] CRISTEA, M., Rev. Univ. "C.I. Parhon" Ploiteh, Bucaresti, 1 (1952), 72.

[20] DERESIEWICZ, H., Proc. 3rd. U.S. National Congress Appl. Mech., 287 (Brown University, Providence, 1958).

[21] ZORSKI, H., Bull. Acad. Polon. Sci. Ser. Sci. Techn. 6 (1958), 331.

[22] STERNBERG, E., Arch. Rational Mech. Anal. 6 (1960), 34.

[23] NOWACKI, W., Bull. Acad. Polon. Sci. Ser. Sci. Techn. 15 (1967), 583.

[24] PODSTRIGAN, Y.S., P.M.M. 6 (1960), 215.

[25] NOWACKI, W., Bull. Acad. Polon. Sci. Ser. Sci. Techn. 12 (1964), 465.

[26] RUDIGER, D., Osterr. Ing.-Arch. 18 (1964), 121.

[27] IONESCU-CAZIMIR, V., Bull. Acad. Polon. Sci. Ser. Sci. Techn. 12 (1964), 565.

[28] WEINER, J.H., Quart. Appl. Math. 15 (1957), 102.

[29] KNOPS, R.J., & PAYNE, L.E., Int. J. Solids Structures 6 (1970), 1173.

[30] BRUN, L., J. Mécanique 8 (1969), 167.

[31] DAFERMOS, C.M., Arch. Rational Mech. Anal. 29 (1968), 241.

[32] IESAN, D., Analele Stiint. Univ. "A.I. Cuza" Iasi, Sect. I. Matematica, 12 (1966), 439.

[33] NICKELL, R.E. & SACKMAN, J.L., Quart. Appl. Math. 26 (1968), 11.

[34] RAFALSKI, P., Int. J. Engng. Sci., 6 (1968), 465.

[35] LESSEN, M., Quart. Appl. Math. 15 (1957), 105.

[36] DERESIEWICZ, H., J. Acoust. Soc. Am. 29 (1957), 204.

[37] CHADWICK, P. & SNEDDON, I.N., J. Mech. Phys. Solids 6 (1958), 223.

[38] CHADWICK, P. & POWDRILL, B., Int. J. Engng. Sci. 3 (1965), 561.

[39] SNEDDON, I.N., The Linear Theory of Thermoelasticity, CISM Courses and Lectures No. 119 (Springer, Wien, 1974).

[40] SNEDDON, I.N., Proc. Roy. Soc. Edinburgh A65 (1958), 121.

[41] EASON, G. & SNEDDON, I.N., Proc. Roy. Soc. Edinburgh A65 (1958), 143.

[42] LOCKETT, F.J. & SNEDDON, I.N., Proc. Edinburgh Math. Soc. (ii) 11 (1959), 237.

[43] CHADWICK, P., Progress in Solid Mechanics, vol. 1,p. 265. (North Holland Publishing Co., Amsterdam, 1960).

[44] PARKUS, H., Instationare Warmespannungen (Springer, Wien, 1959).

[45] MEIXNER, J., Arch. Rational Mech. Anal. 39 (1970), 108.

[46] MULLER, I., Arch. Rational Mech. Anal. 34 (1969), 259.

[47] GURTIN, M.E. & PIPKIN, A.C., Arch. Rational Mech. Anal. 31 (1968), 113.

[48] ROETMAN, E.L., Int.J. Engng. Sci. (in course of publication).

[49] FULKS, W. & GUENTHER, R.B., Czech. Math. J. 21 (1971), 683.

THERMAL STRESSES IN ANISOTROPIC BODIES

Witold Nowacki

1. Introduction

Thermoelasticity embraces a wide field of phenomena. It contains the theory of heat conduction and the theory of strain and stresses due to the flow of heat, when coupling of temperature and deformation fields occurs.

The coupling between deformation and temperature fields was first postulated by J.M.C. Duhamel [1], the originator of the theory of thermal stresses who introduced the dilatation term in the equation of thermal conductivity. However, this equation was not well grounded in the thermodynamical sense. Later, an attempt at the thermodynamical justification of this equation was untertaken by Voigt [2] and Jeffreys [3]. However, it was only in 1956 that Biot [4] gave the full justification of the thermal conductivity equation based on the thermodynamics of irreversible processes [5]. Biot also presented the basic methods for solving the thermoelasticity equation as well as formulating a variational approach (for isotropic body).

Thermoelasticity describes a broad range of phenomena. It is a generalisation of the classical theories of elasticity and thermal conductivity, and it is now a fully developed scientific discipline.

The theory of thermoelasticity in homogeneous isotropic bodies has been treated in detail in scientific literature, problems of thermoelasticity in anisotropic bodies however have been dealt with, in only very few publications. This fact is due not only to the mathematical difficulties of the problem, but follows from the lack of wide practical applications. More and more frequently, however, engineering structure contain materials of macroscopically anisotropic structure, (plates, discs, shells, thick-walled pipes etc.) the properties of which (both elastic and thermal) are different in different directions.

In the present paper, which is of a survey character, attention is focused on the foundations of the thermodynamical theory, on the differential equations and the general energy and variational theorems. We shall give here also a review of problems solved so far; the reader interested in details will find them in quoted papers.

2. Fundamental assumptions and relations of linear thermoelasticity

Let a body be at temperature T_0 in an undeformed and unstressed state. This initial state will be called the natural state, in which it is assumed that the entropy of the body is zero. Owing to the action of extermal forces. i.s. body and surface forces, and under the influence of internal heat sources and surface heating or cooling, the body will be subjected to deformation and temperature change.

Displacements $\underline{u}$ will occur in the body and the temperature change can be written as $\theta = T - T_0$, where T is the absolute temperature of a point $\underline{x}$ of the body. The temperature change is accompanied by stresses σ_{ij} and strains ϵ_{ij}. The quantities $\underline{u}$, θ, σ_{ij}, ϵ_{ij} are function of position $\underline{x}$ and time t. We assume that the temperature change $\theta = T - T_0$ accompanying the deformation is small and does not result in significant variations of the elastic and thermal coefficients which will be regarded as independent of T. In addition to the assumption $|\theta/T_0| \ll 1$ we shall assume that second powers and products of the components of strain may be neglected in comparison with the strains ϵ_{ij}. Thus, attention is restricted to the linear regime where the strains are related to the displacements by

$$\epsilon_{ij} = \frac{1}{2}(u_{i,j} + u_{j,i}), \quad i,j = 1,2,3, \tag{2.1}$$

and the strains satisfy the six compatibility relations

$$\epsilon_{ij,k\ell} + \epsilon_{k\ell,ij} - \epsilon_{j\ell,ik} - \epsilon_{ik,j\ell} = 0, \quad i,j,k,\ell = 1,2,3. \tag{2.2}$$

The main task is now to determine constitutive equations relating the components of the stress tensor σ_{ij} with the components of the strain tensor ϵ_{ij} and of the temperature θ. Thermoelastic disturbances cannot be described in terms of classical thermodynamics and we have to use the relations of the thermodynamics of irreversible processes [5][6].

The constitutive equations are deduced from thermodynamical considerations, taking into account the principle of energy and entropy balance [4 – 6]

$$\frac{d}{dt}\int_V \left(U + \frac{1}{2}\rho v_i v_i\right) dV = \int_V X_i v_i\, dV + \int_A p_i v_i\, dA - \int_A q_i n_i\, dA \tag{2.3}$$

$$\int_V \frac{dS}{dt}\, dV = -\int_A \frac{q_i n_i}{T}\, dA + \int_V \Theta\, dV. \tag{2.4}$$

Here U is internal energy, S is the entropy, X_i the components of the body forces, $p_i = \sigma_{ji} n_j$ the components of the stress vector, q_i the components of the vector of heat flux, n_i the components of the normal to the surface A. Further, $v_i = \partial u_i/\partial t$; and the quantity Θ represents the source of entropy, a quantity always positive in a thermodynamically irreversible process.

$$\Theta = -\frac{q_i T_{,i}}{T^2} > 0. \tag{2.5}$$

The terms on the left-hand side of eq. (2.3) represents the rate of increase of the internal and kinetic energies. The first term on the right-hand side is the rate of increase of the work done by the body forces, and the second the rate of increase of the work done by the surface tractions. Finally, the last term of the right-hand side of eq. (2.3) is the entropy acquired by the body by means of thermal conduction. The left-hand side of eq. (2.4) is the rate of increase of the entropy. The first term on the right-hand side of eq. (2.4) represents the exchange of entropy at the surface, and the second term represents the rate of production of entropy due to heat conduction.

Making use of the equations of motion

$$\sigma_{ji,j} + X_i = \rho \ddot{u}_i , \quad i,j = 1,2,3, \tag{2.6}$$

and using the divergence theorem to transform the integrals we arrive at the local relations

$$\dot{U} = \sigma_{ij} \dot{\epsilon}_{ij} - q_{i,i} , \quad \dot{S} = \Theta - \frac{q_{i,i}}{T} + \frac{q_i T_{,i}}{T^2} . \tag{2.7}$$

Introducing the Helmholtz free energy $F = U - ST$ and eliminating the quantity $q_{i,i}$ from eqs. (2.7) we obtain

$$\dot{F} = \frac{\partial F}{\partial \epsilon_{ij}} \dot{\epsilon}_{ij} + \frac{\partial F}{\partial T} \dot{T}$$
$$= \sigma_{ij} \dot{\epsilon}_{ij} - S\dot{T} - T\left(\Theta + \frac{q_i T_{,i}}{T^2}\right) . \tag{2.8}$$

Assuming that the functions Θ , q_i , σ_{ij} do not explicitly depend on the time derivatives of the functions ϵ_{ij} and T we obtain

$$\sigma_{ij} = \frac{\partial F}{\partial \epsilon_{ij}}, \quad S = -\frac{\partial F}{\partial T} , \quad \Theta + \frac{q_i T_{,i}}{T^2} = 0. \tag{2.9}$$

Let us now expand the function $F(\epsilon_{ij}, T)$ into a infinite series in the neighbourhood of the natural state $(0 , T_0)$

$$F(\epsilon_{ij}, T) = \frac{1}{2} C_{ijk\ell} \epsilon_{ij} \epsilon_{k\ell} - \beta_{ij} \epsilon_{ij} \theta + \frac{m}{2} \theta^2 + ... \tag{2.10}$$

From the expansion $F(\epsilon_{ij}, T)$ we retain only the linear and quadratic terms, confining ourselves to linear relations among stresses σ_{ij} strains ϵ_{ij} and temperature change θ.

Let us now ttake advantage of the expressions (2.9)1,2. We obtain

(2.11) $$\sigma_{ij} = c_{ijk\ell}\, \epsilon_{k\ell} - \beta_{ij}\, \theta, \; \beta_{ij} = \beta_{ji},$$

(2.12) $$S = \beta_{ij}\, \epsilon_{ij} + m\theta .$$

In relations (2.11), we identify Hooke's law generalized for thermoelastic problems. There are called the Duhamel – Neumann relations for an anisotropic body. The components $c_{ijk\ell}$, β_{ij} appropriate to the isothermal state play the role of material constants [7]. In the theory of elasticity of an anisotropic body, the following symmetry properties of the tensor $c_{ijk\ell}$ are proved

(2.13) $$c_{ijk\ell} = c_{jik\ell}, \quad c_{ijk\ell} = c_{ij\ell k}, \quad c_{ij\ell k} = c_{\ell k ij} .$$

The first symmetry relation is obtained from the symmetry of the stress tensor σ_{ij}, while the second form the symmetry of strain tensor ϵ_{ij}. The third relation is a consequence of the relations

$$\frac{\partial^2 F}{\partial \epsilon_{ij}\, \partial \epsilon_{k\ell}} = \frac{\partial^2 F}{\partial \epsilon_{k\ell}\, \partial \epsilon_{ij}} \quad \text{or} \quad \frac{\partial \sigma_{ij}}{\partial \epsilon_{k\ell}} = \frac{\partial \sigma_{k\ell}}{\partial \epsilon_{ij}},$$

These relations lead to a reduction in the number of mutually independent constants from 81 to 21 for a body with general anisotropy.

Let us solve the system of eqs. (2.11) for deformations

(2.14) $$\epsilon_{ij} = s_{ijk\ell}\, \sigma_{k\ell} + \alpha_{ij}\, \theta .$$

The quantities $s_{ijk\ell}$ (coefficients of elastic susceptibility) then satisfy the symmetry relations

(2.15) $$s_{ijk\ell} = s_{jik\ell}, \quad s_{ijk\ell} = s_{ij\ell k}, \quad s_{ijk\ell} = s_{k\ell ij} .$$

Consider now a volume element of the anisotropic body free of stresses on its surface. According to (2.14), we obtain for this element

(2.16) $$\epsilon^{\circ}_{ij} = \alpha_{ij}\, \theta .$$

The relation describes the familar physical phenomenon, namely, the proportionality of the element deformation to the increment of temperature θ. The quantities α_{ij} are the coefficients of linear expansion, and it follows from the symmetry of the tensor ϵ_{ij} that α_{ij} is also symmetric. It should be added that the coefficient of volume thermal expression α_{jj} is an invariant.

From relations (2.11) (2.12) and (2.14) we have

$$\left(\frac{\partial\sigma_{ij}}{\partial\epsilon_{k\ell}}\right)_T = C_{ijk\ell}, \qquad \left(\frac{\partial\sigma_{ij}}{\partial T}\right) = -\beta_{ij} = \alpha_{k\ell}\, C_{ijk\ell}, \tag{2.17}$$
$$\left(\frac{\partial\epsilon_{ij}}{\partial T}\right)_\epsilon = \alpha_{ij}, \qquad \left(\frac{\partial S}{\partial\epsilon_{ij}}\right)_T = \beta_{ij}.$$

From the thermodynamical considerations we obtain: $m = C_\epsilon / T_0$ and

$$S = \beta_{ij}\,\epsilon_{ij} + \frac{C_\epsilon}{T_0}\,\theta, \tag{2.18}$$

where C_ϵ is a specific heat related to unit volume at constant deformation. In the expression for entropy, the first term on the right-hand side is due to the coupling of the deformation and the temperature field, the second term express the entropy caused by the heat flow. The purely elastic term does not appear in this expression.

The postulate of the thermodynamics of irreversible processes will be satisfied if $\Theta > 0$ i.e. when $-q_i T_{,i}/T^2 > 0$. This condition is satisfied by the Fourier law of heat conduction [5]

$$-q_i = k_{ij}\,T_{,j} \quad \text{or} \quad -q_i = k_{ij}\,\theta_{,ij}, \quad \theta = T - T_0. \tag{2.19}$$

It remains to relate the entropy to the thermal conductivity. Combining the relations (2.19) and

$$T\dot{S} = -\operatorname{div}\underline{q} = -q_{i,i} \tag{2.20}$$

we arrive at the equation

$$T\dot{S} = k_{ij}\,\theta_{,ij} \tag{2.21}$$

From the relations (2.18) and (2.21) we obtain

$$\lambda_{ij}\,T_{,ij} = T\beta_{ij}\,\dot{\epsilon}_{ij} + \frac{C_\epsilon}{T_0}\,T\dot{\theta}, \qquad \theta = T - T_0. \tag{2.22}$$

We note that the terms on the right-hand side of this equation make it nonlinear. Putting $T = T_0$ on the right-hand side of (2.22) to linearize the equation gives

$$\lambda_{ij}\,\theta_{,ij} - c_\epsilon \dot{\theta} - T_o\beta_{ij}\,\dot{\epsilon}_{ij} = 0 \tag{2.23}$$

In this extended equation of thermal conductivity, the term $T_o\beta_{ij}\,\dot{\epsilon}_{ij}$ characterizes the coupling of the deformation and temperature fields. If there are internal sources in the body, we should add to (2.23) the quantity W, which determines the amount of heat produced per unit volume and time

$$\lambda_{ij}\,\theta_{,ij} - c_\epsilon \dot{\theta} - T_o\beta_{ij}\,\dot{\epsilon}_{ij} + W = 0. \tag{2.24}$$

The full set of the differential equations of thermoelasticity comprise the equations of motion and the equations of thermal conductivity. The equations of motion

$$\sigma_{ji,j} + X_i = \rho \ddot{u}_i\,(\underline{x},t), \quad \underline{x} \in V, \ t > 0; \tag{2.25}$$

can be transformed, making use of the Duhamel – Neumann equations

$$\sigma_{ij} = C_{ijk\ell}\,\epsilon_{k\ell} - \beta_{ij}\,\theta\,, \tag{2.26}$$

and the strain – displacement relations

$$\epsilon_{ij} = \frac{1}{2}\,(u_{i,j} + u_{j,i}) \tag{2.27}$$

into three equations containing displacements u_i and temperature θ as unknown function

$$C_{ijk\ell}\,u_{k,\ell j} + X_i = \rho\ddot{u}_i + \beta_{ij}\,\theta_{,j}\,, \quad \underline{x} \in V, \quad t > 0. \tag{2.28}$$

The above equations and those of thermal conductivity

$$\lambda_{ij}\,\theta_{,ij} - c_\epsilon \dot{\theta} - T_o\beta_{ij}\,\dot{\epsilon}_{ij} + W = 0 \tag{2.29}$$

are coupled. Body forces, heat sources, heating and heat flow through the surface result in variations of both displacements and temperature. Boundary conditions of a mechanical type are given in the form of displacements u_i or loadings p_i on the surface A. Thermal conditions can be, in a general way, written in the form

$$\alpha \frac{\partial\theta}{\partial n} + \beta\theta = f(\underline{x},t), \quad \underline{x} \in A, \ t > 0, \qquad \alpha,\beta = \text{const.}, \tag{2.30}$$

determining the heat flow through the surface A. The initial conditions e.q. for $t = 0$, are that the displacement u_i, the velocity of these displacements and temperature are the known functions

$$u_i(\underline{x},0) = f_i(\underline{x}), \quad \dot{u}_i(\underline{x},0) = g(\underline{x}), \quad \theta(\underline{x},0) = h(\underline{x}), \ \underline{x} \in V, \ t = 0. \tag{2.31}$$

If the variation of body forces, heat sources and heatings is slow, then the inertia terms in the equations of motion can be omitted and the problem can be regarded as quasi-static. The quasi-static equations of thermoelasticity given below are, however, coupled

$$C_{ijk\ell} \, u_{k,\ell j} + X_i = \beta_{ij} \, \theta_{,j} \, , \tag{2.32}$$

$$\lambda_{ij} \, \theta_{,ij} - C_\epsilon \dot{\theta} - T_0 \beta_{ij} \, \dot{\epsilon}_{ij} + W = 0 \tag{2.33}$$

Thermoelasticity embraces the following subject previously developed separately: classical elastokinetics and the theories of thermal conduction and thermal stress. We shall determine the differential equations of classical elastokinetics assuming that the motion is adiabatic, i.e. without heat exchange in the body. Since for an adiabatic process $\dot{S} = 0$, eq. (2.18) yields $\dot{\theta} = -T_0 / c_\epsilon \beta_{ij} \, \dot{\epsilon}_{ij}$ or, after integrating assuming homogeneous initial conditions:

$$\theta = -\frac{T_o}{c_\epsilon} \beta_{ij} \, \epsilon_{ij} . \tag{2.34}$$

This equation replaces the equation of heat conduction. Inserting (2.34) in (2.28), we obtain the displacements equation of classical elastokinetics

$$(c_{ijk\ell})_s \, u_{k\ell j} + X_i = \rho \ddot{u}_i \, , \tag{2.35}$$

where

$$(C_{ijk\ell})_s = (C_{ijk\ell})_T + \frac{(\beta_{ij})_T \, (\beta_{k\ell})_T}{c_\epsilon} T_o$$

The quantities $(C_{ijk\ell})_s$ are mechanical constants, measured in adiabatic conditions. The constitutive equations, after substituting (2.34) into (2.11) and (2.14), take the form

$$\sigma_{ij} = (C_{ijk\ell})_s \, \epsilon_{k\ell} \, , \quad \epsilon_{ij} = (s_{ijk\ell})_s \, \sigma_{k\ell} \, , \tag{2.36}$$

where

$$(s_{ijk\ell})_s = (s_{ijk\ell})_T - \frac{\alpha_{ij} \, \alpha_{k\ell}}{C_\sigma} T_o \, , \quad C_\sigma = C_\epsilon + \alpha_{ij} \, \beta_{ij} \, T_o .$$

In the theory of thermal stresses which considers the influence of surface and internal heating on the deformation and stresses the term $\beta_{ij}\,\dot{\epsilon}_{ij}$ appearing in the thermal conductivity equation is assumed to be negligible. This simplification leads to the following two independent equations

$$C_{ijk\ell}\, u_{k,\ell j} + X_i = \rho \ddot{u}_i + \beta_{ij}\, \theta_{,j} \tag{2.37}$$

$$\lambda_{ij}\, \theta_{,ij} - C_e \dot{\theta} + W = 0 \tag{2.38}$$

The temperature θ is determined from (2.38) i.e. from the classical equation of thermal conductivity. When we know the temperature distribution, we are able to determine the displacements from eq. (2.37).

In the case of stationary heat flow, the production of entropy is compensated by the exchange of entropy with the environment. This exchange is negative and its absolute value is equal to the entropy production in the body. In the equations of thermoelasticity (2.38) (2.29) the derivatives with respect to time disappear and eq. (2.28) becomes

$$C_{ijk\ell}\, u_{k,\ell j} + X_i = \beta_{ij}\, \theta_{,j}\ . \tag{2.39}$$

The temperature θ appearing in these equations is a know function, obtained by solving the heat conduction equation in the case of a stationary flow of heat

$$\lambda_{ij}\, \theta_{,ij} + W = 0. \tag{2.40}$$

3. The fundamental theorems for anisotropic bodies.

An important part is played in classical elasticity by variational theorems, which consider either variation of the deformation state or variation of the stress state. In what follows we shall present the thermoelastic (coupled) variational theorem for the deformation state, devised by M.A. Biot [4]. This theorem consist of two parts, the first of which utilizes the familiar d'Alembert principle of the elasticity theory

$$\int_V (X_i - \rho \ddot{u}_i)\, \delta u_i\, dV + \int_A p_i\, \delta u_i\, dA = \int_V \sigma_{ij}\, \delta \epsilon_{ij}\, dV. \tag{3.1}$$

In this equation δu_i are the virtual increments of displacement and $\delta\epsilon_{ij}$ the virtual increments of deformation, each assumed to be arbitrary continuous function, and complying with the conditions contraining the body motion. Supplementing the eq. (3.1) with the constitutive equations (2.11), we obtain

$$\int_V (X_i - \rho \ddot{u}_i)\delta u_i \, dV + \int_A p_i \, \delta u_i \, dA = \delta W - \int_V \beta_{ij} \, \delta\epsilon_{ij} \, \theta dV \, , \qquad (3.2)$$

where

$$W = \frac{1}{2} \int_V C_{ijk\ell} \, \epsilon_{ij} \, \epsilon_{k\ell} \, dV. \qquad (3.3)$$

The second part of the variational theorem uses the laws governing the heat flow so we utilize the expressions interrelating heat flow, temperature and entropy

$$q_i = - \lambda_{ij} \, \theta_{,j}, \; T_o \dot{S} \cong - q_{i,i} = \lambda_{ij} \, \theta_{,ij}, \; \dot{S}T_o = \beta_{ij} \, \dot{\epsilon}_{ij} \, T_o + C_\epsilon \dot{\theta}. \qquad (3.4)$$

These relations can be written more conveniently by introducing the vector function $\underline{H}$, related to entropy and heat flow by

$$S = - H_{i,i} \quad , \quad q_i = T_o \dot{H}_i \; . \qquad (3.5)$$

Since

$$q_i = T_o \dot{H}_i = - k_{ij} \, \theta_{,j} \qquad (3.6)$$

and solving these relations with respect to $\theta_{,i}$, we obtain

$$\theta_{,i} = - T_o \lambda_{ij} \, \dot{H}_j \quad , \qquad (3.7)$$

where λ_{ij} is the matrix inverse to the matrix of the coefficients k_{ij} ; $\lambda_{ij} = [k_{ij}]^{-1}$. Multiplying eqs. (3.7) by δH_i integrating over the volume of the body

$$\int_V (\theta_{,i} + T_o \lambda_{ij} \, \dot{H}_j)\delta H_i \, dV = 0, \qquad (3.8)$$

finally, with use of eq. (3.4)2, we obtain

$$\int_A \theta \delta H_n dA + \frac{C_\epsilon}{T_o} \int_V \theta \delta \theta dV + \beta_{ij} \int_V \theta \delta \epsilon_{ij} \, dV + T_o \int_V \lambda_{ij} \, \dot{H}_j \, \delta H_i \, dV = 0. \qquad (3.9)$$

Introducing the notations

$$P = \frac{C_\epsilon}{2T_o} \int_V \theta^2 dV, \qquad \delta D = T_o \int_V \lambda_{ij} \, \dot{H}_j \, \delta H_i \, dV$$

where the functions P and D are called the thermal potential and the dissipation function respectively, and eliminating the integral $\beta_{ij}\int_V \theta\delta\epsilon_{ij}\,dV$ from the equations (3.2) and (3.9), we finally arrive at the required variational principle

$$\delta(W + P + D) = \delta\mathcal{L} - \int_A \theta\delta H_n\, dA, \tag{3.10}$$

where

$$\delta\mathcal{L} = \int_V (X_i - \rho\ddot{u}_i)\delta u_i\, dV + \int_A p_i\,\delta u_i\, dA. \tag{3.11}$$

This is the principle of virtual work for variations of the displacements and the temperature, extended to the problem of thermoelasticity of anisotropic bodies.

a) Consider the particular case when we assume $\theta = -T_0/C_\epsilon\beta_{ij}\,\epsilon_{ij}$, corresponding to the assumption of an adiabatic process.
Then (3.2) transforms into

$$\int_V (X_i - \rho\ddot{u}_i)\delta u_i\, dV + \int_A p\delta u_i\, dA = \delta W_s \tag{3.12}$$

$$W_s = \int_V (C_{ijk\ell})_s\,\epsilon_{ij}\,\epsilon_{k\ell}\, dV.$$

Equation (3.12) constitutes d'Alembert's principle for classical elastokinetics.

b) In the theory of thermal stresses the coupling between the strain and temperature fields is neglected, by neglecting in eq. (2.23) the term $T_0\beta_{ij}\,\dot{\epsilon}_{ij}$. Thus we arrive at two independent variational equations

$$\int_V (X_i - \rho\ddot{u}_i)\delta u_i\, dV + \int_A p_i\,\delta u_i\, dA + \beta_{ij}\int_V \theta\delta\epsilon_{ij}\, dV = \delta W, \tag{3.13}$$

$$\delta(P + D) + \int_A \theta\delta H_n\, dA = 0 \tag{3.14}$$

Eq. (3.14) constitutes variational theorem for the classical heat conduction problem. In eq. (3.13) the temperature is regarded as known, calculated by means of the classical heat conduction equation.

Let us return to the general variational theorem of thermoelasticity, eq. (3.10) and assume that the variational increments δu_i, $\delta\epsilon_{ij}$, δH_i e.t.c. coincide with the increment occurring when the process passes from a time instant t to t + dt. Then

$$\delta u_i = \frac{\partial u_i}{\partial t}\,dt = v_i\,dt, \quad \delta H_i = \dot{H}_i\,dt = -\frac{k_{ij}}{T_0}\,\theta_{,j}\,, \quad \delta W = \dot{W}dt, \text{ e.t.c.}$$

From (3.10) we obtain

$$\frac{d}{dt}(W+K+P) + X_T = \int_V X_i v_i \, dV + \int_A p_i v_i \, dA + \frac{1}{T_0}\int_V k_{ij} \, \theta\theta_{,i} n_j \, dA,$$

where

$$X_T = \frac{1}{T_0}\int_V k_{ij} \, \theta_{,i} \theta_{,j} \, dV, \qquad K = \frac{\rho}{2}\int_V v_i v_i \, dV.$$

Observe that the right-hand side contains the causes producing the motion of the body, i.e. the body forces X_i, surface tractions p_i and the thermal boundary data. Furthermore the integrand $k_{ij} \theta_{,i} \theta_{,j}$ appearing in the function X_i is a positive definite quadratic function.

The energy theorem is used in proving the uniqueness of the solution of the fundamental thermoelasticity differential equations. The proof of uniqueness has been given by V. Ionescu Cazimir [8] for the boundary conditions in displacements or loadings on A.

The proof of uniqueness can also be extended to the case when on a part of the boundary A_σ tractions are given, while on the remaining part displacements u_i are known. Similarly, different thermal conditions can be taken on A_σ and A_u. A detailed exposition of this case as well as another one concerning a discontinuity of stresses on the surface A_σ inside the region V + A is given in the paper by V. Ionescu – Cazimir [8].

One on must interesting theorems of thermoelasticity is the reciprocity theorem. The extended reciprocity theorem is thermoelasticity has been formulated by V. Ionescu – Cazimir [9].

Consider two systems of causes and effects. In the first there act the body forces X_i, surface tractions p_i, heat sources W and surface heatings h. These causes produce in the anisotropic body the displacement u_i and the temperature θ. The causes and effects on the second system will be denoted by "primes".

We base on the Duhamel – Neumann relations, to which the Laplace transform has been applied. From these relations we obtain

$$\int_V (\bar{\sigma}_{ij} + \beta_{ij}\bar{\theta})\bar{\epsilon}'_{ij} \, dV = \int_V (\bar{\sigma}'_{ij} + \beta_{ij}\bar{\theta}')\bar{\epsilon}_{ij} \, dV. \tag{3.16}$$

With the help of the equation of motion we transfrom eq. (3.16) to the form

$$\int_V (\bar{X}_i \bar{u}'_i - \bar{X}'_i \bar{u}_i) dV + \int_A (\bar{p}_i \bar{u}'_i - \bar{p}'_i \bar{u}_i) dA + \beta_{ij}\int_V (\bar{\epsilon}'_{ij}\bar{\theta} - \bar{\epsilon}_{ij}\bar{\theta}') dV = 0. \tag{3.17}$$

We assumed here the homogeneous initial conditions. Eq. (3.17) constitutes the first part of the reciprocity theorem. The second part is derived on the basis of the heat conduction equation. We obtain

$$\int_V k_{ij}(\bar{\theta}_{,ij}\,\bar{\theta}' - \bar{\theta}'_{,ij}\,\bar{\theta})dV - T_o p\int_V \beta_{ij}(\bar{\epsilon}_{ij}\,\bar{\theta}' - \bar{\epsilon}'_{ij}\,\bar{\theta})dV + \int_V (\bar{W}\bar{\theta}' - \bar{W}'\bar{\theta})dV = 0 \qquad (3.18')$$

or

$$-T_o p\int_V \beta_{ij}(\bar{\epsilon}_{ij}\,\bar{\theta}' - \bar{\epsilon}'_{ij}\,\bar{\theta})dV + \int_V (\bar{W}\bar{\theta}' - \bar{W}'\bar{\theta})dV + k_{ij}\int_A (\bar{h}'\bar{\theta}_{,i}\,\bar{h}\bar{\theta}'_{,i})n_j\,dA = 0 \qquad (3.18'')$$

We assumed here the homogeneous initial conditions for temperature and the following boundary conditions

$$(3.19')\quad \theta(\underline{x},t) = h(\underline{x},t), \qquad \theta'(\underline{x},t) = h'(\underline{x},t), \quad \underline{x}\in A, \quad t > 0.$$

Eq. (3.18) constitutes the second part of the reciprocity theorem. If we eliminate from eqs. (3.17) and (3.18") the common term, we arrive after inverting the Laplace transform at an equation containing all causes and effects. We have

$$(3.19'')\quad T_o\int_A (p_i \ominus u'_i - p'_i \ominus u_i)dA + T_o\int_V (X_i \ominus u'_i - X'_i \ominus u_i)dV(\underline{x}) = \int_V (W * \theta' - W' * \theta)dV + \int_A (h' * \theta_{,i} - h * \theta'_{,i})dA,$$

where

$$p_i \ominus u'_i = \int_{t_o}^{t} p_i(\underline{x}, t-\tau)\,\frac{\partial u_i(\underline{x},\tau)}{\partial \tau}\,d\tau,$$

$$W * \theta' = \int_0^t W(\underline{x}, t-\tau)\theta'(\underline{x},\tau)d\tau, \quad \text{e.t.c.}$$

Consider now particular cases of the reciprocity theorem.

a) Assuming that there are no heat sources and no heat exchange between volume elements. We have for this adiabatic process

$$(3.20)\qquad \theta = -\frac{T_o}{C_\epsilon}\beta_{ij}\,\epsilon_{ij}, \qquad \theta' = -\frac{T_o}{C_\epsilon}\beta_{ij}\,\epsilon'_{ij}.$$

We obtain the reciprocity theorem in the form

$$(3.21)\qquad \int_A (p_i * u'_i - p'_i * u_i)dA + \int_V (X_i * u'_i - X'_i * u_i)dV = 0.$$

b) If we treat the dynamical thermoelasticity problem as uncoupled and apply the approximate theory of thermal stresses (neglecting the term $\beta_{ij}\dot{\epsilon}_{ij}$, in the heat conduction equation!). We obtain

$$\int_A (p_i * u_i' - p_i' * u_i)\,dA + \int_V (X_i * u_i' - X_i' * u_i)\,dV + \\ + \beta_{ij} \int_V (\theta * \epsilon_{ij}' - \theta' * \epsilon_{ij})\,dV = 0 \qquad (3.22)$$

$$\int_V (W * \theta' - W' * \theta)\,dV + k_{ij} \int_A (h' * \theta_{,i} - h * \theta'_{,i})\,n_j\,dA = 0. \qquad (3.23)$$

c)In the case of the static problem with a stationary temperature field the reciprocity theorem takes the simpler form [10].

$$\int_A (p_i u_i' - p_i' u_i)\,dA + \int_V (X_i u_i' - X_i' u_i)\,dV + \beta_{ij} \int_V (\theta \epsilon_{ij}' - \theta' \epsilon_{ij})\,dV = 0 \qquad (3.24)$$

$$\int_V (W \cdot \theta' - W' \theta)\,dV + k_{ij} \int_A (h' \theta_{,i} - h \theta'_{,i})\,n_j\,dA = 0. \qquad (3.25)$$

d)Consider finally the particular case of eq. (3.24) when

$$X_i = X_i' = 0, \quad p_i = 0, \quad p_i' = 1 n_i, \quad \sigma_{ji}' = 1\delta_{ij}.$$

We assume that the body is simply-connected, free of body force and its deformation is produced by heating and an action of heat sources. In the primed system we assume the isothermal state and homogeneous extension of the body. Eq. (3.24) is then considerably simplified, namely

$$\int_A n_i u_i\,dA = \beta_{ij} \int_V \theta \epsilon_{ij}'\,dV. \qquad (3.26)$$

The first integral represents the volume increment of the body. Therefore [10]

$$\Delta V = \alpha_{ij} \int_V \theta \sigma_{ij}'\,dV$$

or

$$\Delta V = \alpha_{jj} \int_V \theta\,dV. \qquad (3.27)$$

The volume increment is equal to the integral of the temperature over the region V, multiplied by the invariant α_{jj}. From (2.14) and (3.27) we obtain

$$\Delta V = \int_V \epsilon_{jj}\,dV = \alpha_{jj} \int_V \theta\,dV + s_{jjk\ell} \int_V \sigma_{k\ell}\,dV \qquad (3.28)$$

or

$$s_{jjk\ell} \int_V \sigma_{k\ell}\,dV = 0$$

For the isotropic body, we have [10]

$$\Delta V = 3\alpha_t \int_V \theta dV, \qquad \int_V \sigma_{kk} dV = 0 \tag{3.29}$$

e). Let the anisotropic body contained within the region V and bounded by the surface A be subjected to heating. Let on the part A_u of the surface A, equal to zero, appear displacements u_i and on the part A_σ of the surface A, equal to zero, appear tractions p. Moreover let $X_i = 0$.

For the determining the displacement $u_i(\underline{x}), \underline{x} \in V$ consider a body of the same shape and with the same boundary conditions.

In this body let $\theta' = 0$ and let a concentrated force $X_i' = \delta(\underline{x} - \underline{\xi})\delta_{ik}$ be acting at the point $\underline{\xi}$ which is, consequently directed along the axis x_k.

This force will cause displacements $u_i' = U_i^{(k)}(\underline{x}, \underline{\xi})$ assuming that the functions $U_i^{(k)}(\underline{x}, \underline{\xi})$ are selected such as to satisfy homogeneous boundary conditions on A_σ and A_u.

Making use of formula (3.24) we obtain

$$-\int_V \delta(\underline{x} - \underline{\xi})\delta_{ik} u_i(\underline{x}) dV(\underline{x}) + \beta_{ij} \int_V \theta(\underline{x}) \epsilon_{ij}^{(k)}(\underline{x}, \underline{\xi}) dV(\underline{x}) = 0. \tag{3.30}$$

As a result we obtain the following formula

$$u_k(\underline{\xi}) = \frac{1}{2}\beta_{ij} \int_V \theta(\underline{x}) \left[\frac{\partial U_i^{(k)}(\underline{x}, \underline{\xi})}{\partial x_j} + \frac{\partial U_j^{(k)}(\underline{x}, \underline{\xi})}{\partial x_i} \right] dV(\underline{x}). \tag{3.31}$$

The formula (3.31) may be treated as a generalization of Maysel's formula [11]. For the isotropic body, we obtain

$$u_k(\underline{\xi}) = \gamma \int_V \theta(\underline{x}) \frac{\partial U_j^{(k)}(\underline{x}, \underline{\xi})}{\partial x_j} dV(\underline{x}). \tag{3.32}$$

Here $U_{j,j}^{(k)}(\underline{x}, \underline{\xi})$ should be treated as a dilatation caused at the point $\underline{\xi}$ by a concentrated force X_i applied at the point $\underline{x}$.

5. Three-dimensional problems in anisotropic thermoelasticity

Let us consider the non-coupled quasi-static problem of thermoelasticity. We write the equation of heat conduction

$$\lambda_{ij}\theta_{,ij} - c_\epsilon \dot{\theta} = -W, \tag{4.1}$$

and the displacement equations

$$C_{ijk\ell}\, u_{k,\ell j} = \beta_{ij}\, \theta_{,j} \tag{4.2}$$

in the following form

$$L_{ij}\, u_j = - W\delta_{4i} \qquad i = 1,2,3,4, \qquad u_4 = \theta. \tag{4.3}$$

The L_{ij} are certain linear differential operators of first and second order. The operators $L_{ij} = L_{ji}$ $(i,j = 1,2,3)$ are associated with he displacements equation. We have further

$$L_{i4} = -\beta_{ij}\partial_j\,, \quad L_{4i} = 0, \quad L_{44} = \lambda_{ij}\partial_i\partial_j - c_\epsilon\partial_t\,, \quad i = 1,2,3.$$

Let us express the function u_i by means of four functions χ_i $(i = 1,2,3,4)$ as follows [12]:

$$u_1 = \begin{vmatrix} \chi_1 & L_{12} & L_{13} & L_{14} \\ \chi_2 & L_{22} & L_{23} & L_{24} \\ \chi_3 & L_{32} & L_{33} & L_{34} \\ \chi_4 & 0 & 0 & L_{44} \end{vmatrix}, \quad u_2 = \begin{vmatrix} L_{11} & \chi_1 & L_{13} & L_{14} \\ L_{21} & \chi_2 & L_{23} & L_{24} \\ L_{31} & \chi_3 & L_{33} & L_{34} \\ 0 & \chi_4 & 0 & L_{44} \end{vmatrix}, \ \text{e.t.c.} \tag{4.4}$$

The functions χ_i should satisfy the equations

$$\begin{vmatrix} L_{11} & L_{12} & L_{13} & L_{14} \\ L_{21} & L_{22} & L_{23} & L_{24} \\ L_{31} & L_{32} & L_{33} & L_{34} \\ 0 & 0 & 0 & L_{44} \end{vmatrix} \chi_i = - W\delta_{i4}\,, \tag{4.5'}$$

or

$$\| L_{ij} \| \chi_i = - W\delta_{i4}\,, \quad i,j = 1,2,3,4. \tag{4.5''}$$

The functions χ_i can be regarded as Galerkin's functions, generalized to the case of anisotropic thermoelasticity. The Galerkin method has been used to solving some thermoelastic problem in a simply anisotropic body, the transversely isotropic body [13].

The system of coordinates will be assumed in such a way that the three planes coincide with those of elastic symmetry. Denote by E, ν Young's modulus and Poisson's ratio in the direction x_1, x_2 and by E', ν' the same quantities for the direction x_3. Let α λ denote the coefficients of thermal expansion and heat transfer, respectively in the direction x_1, x_2 and α', λ' the same quantities in the direction x_3. The constitutive equations have the form (*)

$$
(4.6)\qquad
\begin{aligned}
\sigma_{11} &= c_{11}\epsilon_{11} + c_{12}\epsilon_{22} + c_{13}\epsilon_{33} - \beta\theta, & \sigma_{23} &= 2c_{44}\epsilon_{23},\\
\sigma_{22} &= c_{12}\epsilon_{11} + c_{11}\epsilon_{22} + c_{13}\epsilon_{33} - \beta\theta, & \sigma_{31} &= 2c_{44}\epsilon_{31},\\
\sigma_{33} &= c_{13}\epsilon_{11} + c_{13}\epsilon_{22} + c_{33}\epsilon_{33} - \beta'\theta, & \sigma_{12} &= 2c_{66}\epsilon_{12},
\end{aligned}
$$

where the material constants $c_{11},\ldots, c_{44}$ are expressed throught the four quantities E, E', ν, ν' ; and β, β' through the quantities E, E', α, α', ν, ν'. Introducing (4.6) into the equilibrium equations, we obtain the system of equations (4.5"). The quantities L_{ij} have the form

$$
(4.7)\qquad
\left\{
\begin{aligned}
&L_{11} = c_{11}\partial_1^2 + c_{66}\partial_2^2 + c_{44}\partial_3^2,\\
&L_{22} = c_{66}\partial_1^2 + c_{11}\partial_2^2 + c_{44}\partial_3^2,\\
&L_{33} = c_{44}(\partial_1^2 + \partial_2^2) + c_{33}\partial_3^2,\\
&L_{23} = L_{32} = (c_{13} + c_{44})\partial_2\partial_3,\\
&L_{31} = L_{13} = (c_{13} + c_{44})\partial_1\partial_3,\\
&L_{12} = L_{21} = (c_{12} + c_{66})\partial_1\partial_2,\\
&L_{14} = -\beta\partial_1,\quad L_{24} = -\beta\partial_2,\quad L_{34} = -\beta'\partial_3,\\
&L_{41} = L_{42} = L_{43} = 0,\quad L_{44} = \lambda(\partial_1^2 + \partial_2^2) + \lambda'\partial_3^2 - c_e\partial_t .
\end{aligned}
\right.
$$

Introducing (4.7) into (4.5') and performing the operations indicated in (4.5) we can represent that equation in the form

$$
(4.8)\qquad
\begin{aligned}
&\lambda' c_{33} c_{44}^2 (\mu_1^2\nabla_1^2 + \partial_3^2)(\mu_3^2\nabla_1^2 + \partial_3^2)\ \times\\
&\times(\mu_5^2\nabla_1^2 + \partial_3^2 - \sigma^2\partial_t)(\mu_7^2\nabla_1^2 + \partial_3^2)X_i = -W\delta_{4i}
\end{aligned}
\qquad i = 1,2,3,4,
$$

(*) The linear elasticities are components of a positive definite fourth-order tensor. Necessary and sufficient conditions for the satisfaction of the latter requirement are

$$c_{11} > 0,\ c_{11} > c_{12},\ c_{11}^2 > c_{12}^2,\ c_{44} > 0,\quad c_{33}(c_{11} + c_{12}) > 2c_{13}^2$$

where

$$\mu_{1,3}^2 = \begin{cases} \epsilon^2\left(\rho \pm \sqrt{\rho^2 - 1}\right) & \text{for } \rho > 1 , \\ \epsilon^2 & \text{for } \rho = 1 , \\ \epsilon^2\left(\sqrt{\frac{1+\rho}{2}} \pm i \sqrt{\frac{1-\rho}{2}}\right)^2 & \text{for } \rho < 1 , \end{cases} \tag{4.9}$$

$$\mu_5^2 = \frac{\lambda}{\lambda'} , \quad \mu_7^2 = \frac{c_{66}}{c_{44}} , \quad \epsilon^4 = \frac{c_{11}}{c_{33}} , \quad \sigma^2 = \frac{C_\epsilon}{\lambda'},$$

$$\rho = \frac{c_{11}\, c_{33} - 2c_{13}\, c_{44} - c_{13}^2}{2c_{44}\,(c_{11}\, c_{33})^{1/2}} , \quad \nabla_1^2 = \partial_1^2 + \partial_2^2 .$$

For the solution of solved problems, two function χ_3 and χ_4 were sufficient. Since in the expressions for u_i , the operator $\mu_7^2 \nabla_1^2 + \partial_3^2$ will appear, therefore we assume that

$$\begin{cases} \varphi = c_{44}\,(\mu_7^2 \nabla_1^2 + \partial_3^2)(\mu_3^2 \nabla_1^2 + \partial_3^2 - \sigma^2 \partial_t)\chi_3 , \\ \psi = c_{44}\,(\mu_7^2 \nabla_1^2 + \partial_3^2)\chi_4 . \end{cases} \tag{4.10}$$

The functions ψ and φ satisfies the equations

$$\lambda' c_{33}\, c_{44}\,(\mu_1^2 \nabla_1^2 + \partial_3^2)(\mu_3^2 \nabla_1^2 + \partial_3^2)(\mu_5^2 \nabla_1^2 + \partial_3^2 - \sigma^2 \partial_t)\psi = - W , \tag{4.11}$$

$$(\mu_1^2 \nabla_1^2 + \partial_3^2)(\mu_3^2 \nabla_1^2 + \partial_3^2)\varphi = 0. \tag{4.12}$$

It is easy to see that the function φ is the Galerkin function generalized to the case of transverse isotropy [14 – 15].

In the case of axially symmetric problems, it will be more convenient to use displacements and stresses expressed in cylindrical coordinates. Introducing the notation $\nabla_r^2 = \partial^2/\partial r^2 + 1/r\, \partial/\partial r$ we obtain the differential equations for the function φ and ψ, substituting in the place of ∇_1^2 in the Eqs. (4.10) (4.11) the operator ∇_r^2 .

The displacements and the temperature are given by the equations

$$u_r = \beta c_{44} \frac{\partial}{\partial r} (\nabla_r^2 + a\partial_z^2)\psi - c_{44}\, s\, \partial_r \partial_z \varphi , \quad u_\varphi = 0 ,$$

$$W = \beta c_{44} \frac{\partial}{\partial z} (b\nabla_r^2 + c\partial_z^2)\psi + c_{44}\,(t\nabla_r^2 + \partial_z^2)\varphi , \tag{4.13}$$

$$\theta = c_{33}\, c_{44}\,(\mu_1^2 \nabla_r^2 + \partial_z^2)(\mu_3^2 \nabla_r^2 + \partial_z^2)\psi , \quad r = (x_1^2 + x_2^2)^{1/2}, \; z = x_3 ,$$

where

$$a = \eta - \varkappa(1+\gamma\eta), \quad b = \mu_1^2\mu_3^2\varkappa\eta - \gamma\eta - 1, \quad \gamma = \frac{a_{13}}{a_{33}}$$

$$c = \varkappa = \beta'/_{\beta} \quad , \quad s = 1+\gamma\eta \; , \quad t = \mu_1^3\mu_3^2\eta, \quad \eta = \frac{\mu_1^2+\mu_3^2+2\gamma}{\mu_1^2\mu_3^2-\gamma^2} = \frac{a_{33}}{a_{44}} ,$$

The assumption of the functions φ and ψ suffices for the determination of thermal stresses in a simple system: an infinite space, a semi-space and an elastic layer. On the other-hand, for the solution of the state of stress in thick circular and rectangular plates, we should, to satisfy all the boundary conditions, take for the solution, besides the function φ and ψ, also the functions χ_1 and χ_2. With the help of the two functions φ and ψ a few problems have been solved.

The problem was solved in closed form derived for thermal stresses produced by heat sources in an infinite elastic space and an elastic semi-space, for various static and thermal boundary conditions. Similar solutions were given in the case of an action of a nucleus of thermoelastic strain. The case of stationary heating of an elastic semi space and a layer was also examined. Further it was proved that the stresses the vector of which is perpendicular to the plane bounding the semi-space do not vanish, which was the case in the problem of Sternberg and Mac Dowell [16]. Finally, solutions were given for a few quasi-static problems concerning an action of an instantaneous heat source in an elastic space and a semi-space.

In the case of an infinite elastic solid, to although it is possible to obtain a formal solution of the system of equations (4.1) and (4.2) by applying the quadruple Fourier integral transforms, it has not been possible so far to obtain the solution in a form suitable for calculations of a three-dimensional problem of either general anisotropy or orthogonal anisotropy (orthotropy) [17]. Only in the case of a transversally isotropic body we obtain the results in closed form.

We return to the displacement equations (4.2). A solution of equations (4.2) can be represented in the form of a sum, the first component $\bar{u}_i$ satisfying the non homogeneous system of equations (4.2), while the second component $\bar{\bar{u}}_i$ satisfies the homogeneous system

$$L_{ij}\,\bar{\bar{u}}_j = 0, \quad i = 1,2,3. \tag{4.14}$$

$\bar{\bar{u}}_i$ can be represented in terms of the three functions η_i (i = 1,2,3,) these functions satisfy the homogeneous equation

$$\| L_{ij} \|\, \eta_j = 0, \quad i,j = 1,2,3. \tag{4.15}$$

The particular solution $\bar{u}_i$ for the isotropic body can be devised by the method presented by Goodier [18], who introduced the so-called thermoelastic displacement potential ϕ according to the relation $\bar{u}_i = \phi_{,i}$.

In the case of transversely isotropic body Borş [19] introduced three thermoelastic displacement potentials ϕ_i $(i = 1,2,3)$ according to the relations

$$\bar{u}_1 = \partial_1(\phi_1 + \phi_2) - \partial_2\phi_3, \quad \bar{u}_2 = \partial_2(\phi_1 + \phi_2) + \partial_1\phi_3, \qquad \bar{u}_3 = \partial_3(k_1\phi_1 + k_2\phi_2). \tag{4.16}$$

Introducing (4.16) into the system of equations (4.2), we obtain the following three differential equations

$$(\nabla_1^2 + a_1^2\partial_3^2)\phi_1 = A_1\theta, \quad (\nabla_1^2 + a_2^2\partial_3^2)\phi_2 = A_2\theta, \qquad (\nabla_1^2 + \hat{\beta}^2\partial_3^2)\phi_3 = 0, \quad \hat{\beta}^2 = \frac{c_{44}}{c_{66}}, \quad \nabla_1^2 = \partial_1^2 + \partial_2^2, \tag{4.17}$$

where

$$A_1 = \frac{[c_{13} + c_{44}(1 + k_1)]\beta - c_{11}\beta'}{c_{11}c_{44}(k_1 - k_2)}$$

$$A_2 = \frac{[c_{13} + c_{44}(1 + k_2)]\beta - c_{11}\beta'}{c_{11}c_{44}(k_1 - k_2)} .$$

We obtain the roots k_1, k_2 from the equation

$$c_{44}(c_{13} + c_{44})k^2 + [(c_{13} + c_{44})^2 + c_{44}^2 - c_{11}c_{33}]k + (c_{13} + c_{44})c_{44} = 0. \tag{4.18}$$

The quantities a_1^2, a_2^2 are given by the formulae

$$a_\alpha^2 = \frac{c_{33}k_\alpha}{c_{13} + c_{44}(1 + k_\alpha)} = \frac{c_{44} + (c_{13} + c_{44})k_\alpha}{c_{11}}, \quad \alpha = 1,2. \tag{4.19}$$

For a bounded body the functions ϕ_i $(i=1,2,3)$ satisfy a part of the boundary conditions. Therefore the incomplete solution $\bar{u}_i$ must be completed by a solution $\bar{\bar{u}}_i$ of the system of homogeneous equations (4.14). The functions $\bar{\bar{u}}_i$ must so be chosen that the final solution $u_i = \bar{u}_i + \bar{\bar{u}}_i$ satisfies all the boundary conditions of the problem. The system of equations (4.14) can be solved means of the methods, indicated by Elliot [20], Hu-Hai-Chang [15] and Nowacki [10].

We now give the formule for the displacement $\bar{u}_i$ for the axi-symmetrical case (21). Introducing two displacement potentials $\phi_\alpha (\alpha = 1,2)$, we obtain

$$(4.20) \qquad u_r = \frac{\partial}{\partial r}(\phi_1 + \phi_2), \quad u_z = \frac{\partial}{\partial z}(k_1\phi_1 + k_2\phi_2).$$

Introducing the above relations into the equations (4.2), written for transversally isotropic body in cylindrical coordinate system, we arrive at two equations

$$(4.21) \qquad (\nabla_r^2 + a_1^2\partial_z^2)\phi_1 = B_1\theta, \qquad (\nabla_r^2 + a_2^2\partial_z^2)\phi_2 = B_2\theta,$$

where

$$B_1 = \frac{\beta[(c_{13} + c_{44}) + c_{44}k_2] - \beta' c_{11}}{c_{11}c_{44}(k_2 - k_1)},$$

$$B_2 = -\frac{\beta[(c_{13} + c_{44}) + c_{44}k_1] - \beta' c_{11}}{c_{11}c_{44}(k_2 - k_1)}, \quad \nabla_r^2 = \frac{\partial^2}{\partial r^2} + \frac{1}{r}\frac{\partial}{\partial r}$$

The quantities k_1, k_2 and μ_α^2 ($\alpha = 1,2$) can be calculated by means of the formulae (4.18) (4.19). A number of problems on axisymmetric thermal stresses in a semi-space of transversal isotropy were solved by Artar Singh [22] who employed two displacement functions.

A different methods of solving the problem of stationary and quasi-static thermal stresses in bodies of transversal isotropy have been developed by Grindei [23] and Sharma [24]. Sharma investigated thermal stresses due to the heating of the plane bounding an elastic semi-space; he deduced the solution by introducing two stress functions, which satisfy a differential equation of second order.

5. Plane problems of anisotropic thermoelasticity

Two-dimensional problems have been dealt with fairly extensively. Thus W.H. Pell [27] examined the problem of simultaneous bending and compression of an anisotropic plate, produced by a stationary field varying linearly with the thickness of the plate; in particular he investigated in detail the circular plate.

Mossakowski [22] applied the complex variable method to derive a number of solutions for the action of a heat source in a semi-infinite disc of isogonal anisotropy. It is convenient here to introduce a stress function analogous to Airy's function in isotropic discs [28]. A method of solution for orthotropic discs making use of the type of function of Airy and Marguerré was presented for static and dynamics problems by P.P. Teodorescu [29].

For the orthotropic plate we denote by E_1 and E_2 the Young's moduli in the direction of x_1 and x_2 axis, respectively, by ν–Poisson's ratio and by G the shear modulus. Finally α_1 and α_2 denote the coefficients of thermal expansion and λ_1 , λ_2 coefficients of thermal conductivity in the directions of x_1 and x_2 axes, respectively.

The heat equation for an orthotropic plate has the form

$$(\lambda_1 \partial_1^2 + \lambda_2 \partial_2^2 - C_v \partial_t)\theta = - W. \tag{5.1}$$

The relations between stress, strain and temperature in the plane state of stress are

$$\begin{aligned} \epsilon_{11} &= a_{11}\sigma_{11} + a_{12}\sigma_{12} + \alpha_1\theta, \\ \epsilon_{22} &= a_{21}\sigma_{11} + a_{22}\sigma_{22} + \alpha_2\theta, \\ \epsilon_{12} &= a_{66}\sigma_{12}, \quad a_{11} = \frac{1}{E_1}, \quad a_{22} = \frac{1}{E_2} \quad a_{12} = a_{21} = -\frac{\nu}{E_1}, \quad a_{66} = \frac{1}{2\theta}. \end{aligned} \tag{5.2}$$

Substituting the strains into the compatibility relation

$$\partial_1^2 \epsilon_{22} + \partial_2^2 \epsilon_{11} = 2\partial_1\partial_2\epsilon_{12}, \tag{5.3}$$

and expressing the stresses by means of the Airy function

$$\sigma_{\alpha\beta} = - \partial_\alpha \partial_\beta F + \delta_{\alpha\beta} \nabla^2 F, \quad \alpha,\beta = 1,2 \tag{5.4}$$

we obtain the differential equation

$$\begin{aligned} \varkappa^4 \partial_1^4 F + 2\eta\, \varkappa^2 \partial_1^2 \partial_2^2 F + \partial_2^4 F &= - E_1 (\alpha_1 \partial_2^2 + \alpha_2 \partial_1^2)\theta, \\ \varkappa^4 = \frac{E_1}{E_2}, \quad 2\varkappa^2\eta &= E_1\left(\frac{1}{G} - \frac{2\nu}{E_1}\right). \end{aligned} \tag{5.5}$$

Let us write the solution of eq. (5.5) as the sum of two components $\bar{F}$ and $\bar{\bar{F}}$ where $\bar{F}$ is the particular integral of eq. (5.5) and $\bar{\bar{F}}$ satisfies the homogeneous quasi-biharmonic equation

$$\varkappa^4 \partial_1^4 \bar{\bar{F}} + 2\eta\varkappa^2 \partial_1^2 \partial_2^2 \bar{\bar{F}} + \partial_2^4 \bar{\bar{F}} = 0, \tag{5.6}$$

and the boundary conditions. The described procedure is particularly convenient in the case of boundary conditions expressed in stresses. It is also easy to extend to the case of orthotropic discs the "plate analogy" devised by Dubas [30] and Trommel [31].

If we solve the equations (5.2) with respect to the stresses and introduce into the equilibrium equations, we obtain with (5.1) a system of three equations

$$L_{ij}\, u_j = -W\delta_{i3}\,, \quad i = 1,2,3, \quad u_3 = \theta, \tag{5.7}$$

where

$$\begin{aligned} &L_{11} = c_{11}\,\partial_1^2 + c_{66}\,\partial_2^2\,, \quad L_{22} = c_{66}\,\partial_1^2 + c_{22}\,\partial_2^2\,, \\ &L_{12} = L_{21} = (c_{12} + c_{66})\partial_1\partial_2\,, \quad L_{13} = -\beta_1\partial_1\,, \quad L_{23} = -\beta_2\partial_2\,, \\ &L_{33} = \lambda_1\partial_1^2 + \lambda_2\partial_2^2 - C_e\partial_t\,, \quad L_{31} = L_{32} = 0. \end{aligned} \tag{5.8}$$

Let us express the functions $u_\alpha\,(\alpha = 1,2)$, $u_3 = \theta$ by means of three displacement functions χ_i $(i = 1,2,3)$ as follows

$$u_1 = \begin{vmatrix} \chi_1 & L_{12} & L_{13} \\ \chi_2 & L_{22} & L_{23} \\ \chi_3 & 0 & L_{33} \end{vmatrix}, \quad u_2 = \begin{vmatrix} L_{11} & \chi_1 & L_{13} \\ L_{21} & \chi_2 & L_{23} \\ 0 & \chi_3 & L_{33} \end{vmatrix}, \quad u_3 = \begin{vmatrix} L_{11} & L_{12} & \chi_1 \\ L_{21} & L_{22} & \chi_2 \\ 0 & 0 & \chi_3 \end{vmatrix}. \tag{5.9}$$

The functions χ_i should satisfy the equations

$$\begin{vmatrix} L_{11} & L_{12} & L_{13} \\ L_{21} & L_{22} & L_{23} \\ 0 & 0 & L_{33} \end{vmatrix} \chi_i = -W\delta_{i3}\,, \quad i = 1,2,3, \tag{5.10}$$

or

$$c_{22}\, c_{66}\, (\lambda_1\partial_1^2 + \lambda_2\partial_2^2 - C_e\partial_t)(\mu_1^2\partial_1^2 + \partial_2^2)(\mu_2^2\partial_1^2 + \partial_2^2)\chi_i = -W\delta_{i3}\,, \tag{5.11}$$

where

$$\mu_{1,2}^2 = \varkappa^2 \begin{cases} \sigma \pm \sqrt{\sigma^2 - 1} & \text{for } \sigma > 1, \\ \sigma & \text{for } \sigma = 1, \\ \left(\sqrt{\dfrac{1+\sigma}{2}} \pm i\sqrt{\dfrac{1-\sigma}{2}}\right)^2 & \text{for } \sigma < 1. \end{cases}$$

The functions X_1, X_2 satisfy the homogeneous and the function X_3 the non-homogeneous differential equation. The solution procedure is a follows. From the eq. (5.11) we determine for $i = 3$ the particular integral X_3. By means of the functions X_1, X_2 we satisfy the given boundary conditions.

The plane problem can also be solved by means of two termoelastic displacement potentials [33] ϕ_α ($\alpha = 1,2$).

The displacement equations for a medium of an arbitrary curvilinear orthotropy were derived and examined in the paper of Nowiński, Olszak and Urbanowski [34]; these authors solved three examples, the first of which concerns a non-uniformly heated thick-walled cylinder of cylindrical orthotropy, the second deals with an analogous problem for a disc, and the third as that of the non-uniform heating of a shperical shell of spherical orthotropy.

Interesting investigations on states free of stresses in anisotropic bodies was carried out by Olszak [35]. He proved that for bodies which deform freely and possess rectlinear anisotropy only linear distribution of temperature result in no stresses. However, in the case of bodies of curvilinear anisotropy the compatibility equations constitute much stronger limitation than for bodies with rectilinear anisotropy. For instance, for bodies of spherical orthotropy only a constant distribution of temperature produces no stresses, while for bodies of spherical orthotropy any non-vanishing field result in a state of stress.

Only a few dynamic problems have been solved so far in the literature of the subject. Above all the work by T. Chadwik and L.T.C. Seet [36] deserves notice. The main aim of their work are considerations on the propagation of elastic waves in a transversely isotropic heat-conduction material. The authors are concerned with the system of equations

$$
\begin{aligned}
&\frac{1}{2}(c_{11} - c_{12})\nabla_1^2 u_\alpha + \frac{1}{2}(c_{11} + c_{12}) u_{\beta,\alpha\beta} + c_{44} u_{\alpha,33} + \\
&+ (c_{13} + c_{44}) u_{3,\alpha 3} - \rho\ddot{u}_\alpha = \beta_0 \theta_{,\alpha} \,, \qquad (5.13)\\
&c_{44}\nabla_1^2 u_3 + c_{33} u_{3,33} + (c_{13} + c_{44}) u_{\beta,3\beta} - \rho\ddot{u}_3 = \beta_0' \theta_{,3} \\
&c_\epsilon \dot{\theta} + T_0(\beta_0 \dot{u}_{\alpha,\alpha} + \beta_0' \dot{u}_{3,3}) = \lambda\nabla_1^2\theta + \lambda'\theta_{,33} \,. \quad \alpha,\beta = 1,2
\end{aligned}
$$

This system may be partialy divided through the introduction of three new scalar funcitons ϕ, X, ψ, related to the displacements in the following manner

$$
u_1 = \phi_{,1} + X_{,2} \,, \quad u_2 = \phi_{,2} - X_{,1} \,, \quad u_3 = \psi_{,3} \,. \qquad (5.14)
$$

Setting the above to (5.13), we obtain

$$
\begin{aligned}
& c_{11}\,\nabla_1^2\phi + c_{44}\,\phi_{,33} + (c_{13} + c_{44})\psi_{,33} - \rho\ddot{\phi} = \beta_0\theta\,, \\
(5.15)\qquad & (c_{13} + c_{44})\nabla_1^2\phi + c_{44}\,\nabla_1^2\psi + c_{33}\,\psi_{,33} - \rho\ddot{\psi} = \beta_0'\theta\,, \\
& C_e\dot{\theta} + T_0(\beta_0\nabla_1^2\dot{\phi} + \beta_0'\dot{\psi}_{,33}) = \lambda\nabla_1^2\theta + \lambda'\theta_{,33}\,,
\end{aligned}
$$

as well as

$$
(5.16)\qquad \frac{1}{2}(c_{11} - c_{12})\nabla_1^2\chi + c_{44}\,\chi_{,33} - \rho\ddot{\chi} = 0.
$$

Owing to substitution of (5.1), it was possible to isolate the SH-wave generated by the function χ. It is obvious that only the waves characterized by potentials ϕ and ψ. are coupled with the temperature field.

It is clear here that a transversely isotropic elastic material can transit three body waves in each direction, a quasi-longitudinal wave, a quasi-transverse wave, and a purely transverse wave. The quasi-longitudinal and quasi-transverse waves respectively are modified; both these waves suffer dispersion and attenuation.

In the paper discussed in detail – and exemplified also – the numerical results related to the propagation of plane harmonic waves in a single crystal of zinc. A further expansion of the topic is the paper of the above quoted authors concerning the second-order thermoelasticity theory for isotropic and transversely isotropic materials [37].

REFERENCES

[1] DUHAMEL J.M.C. : Seconde mémoire sur les phénomènes thermomécaniques. J. de l'Ecole Polytechnique 15 (1839), 1 – 15.

[2] VOIGT W. : Lehrbuch der Kristallphysik, Teubner, 1910.

[3] JEFFREYS H. : The thermodynamics of an elastic solid. Proc. Cambr.Phil. Soc., 26 (1930).

[4] BIOT M.A. : Thermoelasticity and irreversible thermodynamics. J. Appl.Phys., 27 (1956).

[5] de GROOT S.R. : Thermodynamics of irreversible processes. Amsterdam 1952.

[6] BOLEY B.A. and WEINER J.H. : Theory of thermal stresses. John Viley, New York, 1960.

[7] NEY J.F. : Physical properties of crystals. Oxford, Clarendon Press, 1957.

[8] IONESCU–CAZIMIR V. : Problem of linear Thermoelasticity. Uniqueness theorems (I) (II), Bull.Acad. Polon.Sci., Série Sci Techn., 12, 12 (1964).

[9] IONESCU–CAZIMIR V. : Problem of linear coupled thermoelasticity, (I) (II). Bull.AcAcad. Polon.Sci., Série Sci. Techn. 9, 12 (1964) and 9, 12 (1964).

[10]NOWACKI W. : Thermal stresses in anisotropic bodies (I). (in Polish), Arch. Mech. Stos. 3, 6 (1954).

[11]MEYSEL V.M. : Temperature problems of the theory of elasticity (in Russian), Kiew, 1951.

[12]MOSIL Gr.C. : Matricile asociate sistemelor de ecuatii cu derivate partiale. Introducere in stidiul cercetazilor lui I.N. Lopatinski. Edit. Acadaemiei R.P.R. Bucuresti, (1950).

[13] MOSSAKOWSKA Z. and NOWACKI W. : Thermal stresses in transversely isotropic bodies Arch. Mech. Stos. 4, 10, (1958).

[14] NOWACKI W. : The determining of stresses and deformation in transversely isotropic elastic bodies. (in Polish). Arch. Mech. Stos. 5, 4 (1953).

[15] HU HAI–CZANG. : On the threedimensional problems of the theory of elasticity of a transversely isotropic body. Acta Sci. Sinica, 2, 2 (1953).

[16] STERNBERG E. and MAC DOWELL E.L. : On the steady-state thermoelastic problem for the half-space. Quart. Appl. Math. (1957).

[17] CARRIER C.P. : The thermal stress and body force problem of infinite orthotropic solid. Quart. Appl. Math., (1944).

[18] GOODIER J.N. : On the integration of the thermoelastic equations. Phil. Mag. VI, 23, (1937), p. 1017.

[19] BORS C.I. : Sur le problème à trois dimensions de la thermoelasticité des corps transversalement isotropes. Bull. de l'Acad. Polon. Sci. Ser. Sci. Tech., 11, 5 (1963), 177-181.

[20] ELLIOT A.H. : The three-dimensional stress distribution in hexagonal aelotropic crystals. Proc. Cambridge Phil. Soc. 44 (1948), 621-630.

[21] BORS C.I. : Tensions axiallement symétriques, dans les corps transversalement isotropes. An. stiint. Univ. Iasi, 8, 1, (1962), 119-126.

[22] SINGH Avtar : Axisymmetrical thermal stresses in transversely isotropic bodies. Arch. Mech. Stos. 12, 3, (1960), 287-304.

[23] GRINDEI I : Tensiuni termice in medii elastice transversal izotrope. An. stiint. Univ. Iasi, 14, 1, (1968), 169-176.

[24] SHARMA B. : Thermal stresses in transversely isotropic semi-infinite elastic solids. J. Appl. Mech. 1958.

[25] IESAN D. : Tensiuni termice in bare ortho ropic. An. stiint. Univ. Iasi, 12, 2 (1967).

[26] IESAN D. : Tensions thermiques dans des barres élastiques non-homogènes. An. stiint. Univ. Iasi, 14, 1 (1968).

[27] PELL W.H. : Thermal deflection of anisotropic thin plates. Quart. Appl. Mech., (1946.).

[28] MOSSAKOWSKI J. : The state of stress and displacemnt in a thin anisotropic plate to a concentrated source of heat. Arch. Mech. Stos. 9, 5 (1957), p. 595.

[29] TEODORESCU P.P. : Asupra problemei plane a elasticitatii unor corpuri anizotrope. VIII. Influenta variatiei de temperatura. Com. Acad. R.P.R. 8, 11 (1958), p. 1119-1126.

[30] TREMMEL F. : Uber die Anwendung der Plattentheorie zur Bestimmung der Wärmespannungsfelder. Osterr. Ing. Arch. 1957.

[31] DUBAS P. : Calcul numérique des plaques et des parois minces, Zürich, 1955.

[32] NOWACKI W. : Thermal stresses in orthotropic plates. Bull. de l'Acaad. Polon. Sci., Ser Sci. Techn. 7, 1 (1959), 1-6.

[33] BORŞ C.I. : Tensioni termice la corpurile orthotrope in cazul problemelor plane. St. si cerc. stiint. Filiala Iasi Acad. R.P.R. 14, 1 (1963), 187-192.

[34] NOWINSKI J., OLSZAK W. and URBANOWSKI W. : On thermoelastic problems in the case of a body of an arbitrary type of curvilinear orthotropy (in polish), Arch. Mech. Stos. 7, 2, (1955), 247-265.

[35] OLSZAK W. : Autocontraints des milieux anisotropes. Bull. Acad. Polon. Sci. Lettr. Cl. Ser. Math. (1950).

[36] CHADWICK P. and SEET L.T.C. : Wave propagation in a transversely isotropic heat-conducting elastic material. Mathematika 17, (1970) p. 255.

[37] CHADWICK P. and SEET L.T.C. : Second-order thermoelasticity theory for isotropic and transversely isotropic materials. Trends in Elasticity and Thermoelasticity. Wolters-Nordhoff Publ., Groningen, 1971.

MAGNETO – AND ELECTROTHERMOELASTICITY

H. PARKUS
Technical University of Vienna

Magneto-electro-thermoelasticity describes the interaction of four fields in an elastic or viscoelastic solid: stress field, displacement field, temperature field and electromagnetic field. There would be additional fields in an oriented medium, but little has been done so far to include effects of this kind in the theory, cf. [1] and [2].

Research in the field started in 1955 with a paper by Knopoff [3]. Since then progress has been steady but relatively slow. An introductory survey has been given by Parkus [4], [5]. The following report links up with [5] and briefly describes what progress has been made during the period 1971 – 1974. Reference to [5] will be made frequently. In particular, the same notation will be used. For convenience, this notation is also given here in the appendix.

1. Electrodynamics of Slowly Moving Bodies

Continuum electrodynamics is governed by the Maxwell equations:

$$\left.\begin{aligned} &\nabla \times \underset{\sim}{H} = \underset{\sim}{j} + \frac{\partial \underset{\sim}{D}}{\partial t}, \qquad \nabla \times \underset{\sim}{E} = -\frac{\partial \underset{\sim}{B}}{\partial t} \\ &\nabla \cdot \underset{\sim}{D} = \rho_e, \qquad \nabla \cdot \underset{\sim}{B} = 0 \end{aligned}\right\} \tag{1.1}$$

These equations are valid both inside and outside of matter. In vacuum we have $\underset{\sim}{j} = 0$, $\rho_e = 0$ and

$$\underset{\sim}{D} = \epsilon_o \underset{\sim}{E}, \quad \underset{\sim}{B} = \mu_o \underset{\sim}{H} \tag{1.2}$$

In a medium at rest, Eqs. (1.2) change to

$$\underset{\sim}{D} = \epsilon_o \underset{\sim}{E} + \underset{\sim}{P}, \quad \underset{\sim}{B} = \mu_o (\underset{\sim}{H} + \underset{\sim}{M}) \tag{1.3}$$

where $\underset{\sim}{P}$ and $\underset{\sim}{M}$ represent polarization and magnetization, respectively, per unit volume. Substitution of Eqs. (1.3) into (1.1) renders

$$\left.\begin{aligned} &\nabla \times \frac{\underset{\sim}{B}}{\mu_o} = \underset{\sim}{j} + \epsilon_o \frac{\partial \underset{\sim}{E}}{\partial t} + \nabla \times \underset{\sim}{M} + \frac{\partial \underset{\sim}{P}}{\partial t}, \qquad \nabla \times \underset{\sim}{E} = -\frac{\partial \underset{\sim}{B}}{\partial t} \\ &\epsilon_o \nabla \cdot \underset{\sim}{E} = \rho_e - \nabla \cdot \underset{\sim}{P}, \qquad \nabla \cdot \underset{\sim}{B} = 0 \end{aligned}\right\} \tag{1.4}$$

There seems to be a general agreement as to the validity of Eqs. (1.4). A controversial subject, however, is the extension of these equations to a moving body(*). In [5] the equations of Minkowski have been used. In recent years there has been a tendency to employ the equations proposed by Chu. These equations differ from those given in Chapter 1 of [5] in that $\underset{\sim}{P}$ is to replaced by $\underset{\sim}{P} + \underset{\sim}{P}_{equ}$ and $\underset{\sim}{E}$ by $\underset{\sim}{E} + \mu_o \underset{\sim}{M} \times \underset{\sim}{v}$, where

$$\underset{\sim}{P}_{equ} = \frac{1}{c^2}(\underset{\sim}{v} \times \underset{\sim}{M}) \tag{1.5}$$

is the "equivalent polarization" for the Ampere moving current loop model. Chu's equations then read, cf. [6], Eqs. (3.40),

$$\left.\begin{aligned}
\nabla \times \underset{\sim}{H} &= \underset{\sim}{j} + \epsilon_o \frac{\partial \underset{\sim}{E}}{\partial t} + \frac{\partial \underset{\sim}{P}}{\partial t} + \nabla \times (\underset{\sim}{P} \times \underset{\sim}{v}) \\
\frac{1}{\mu_o} \nabla \times \underset{\sim}{E} &= -\frac{\partial \underset{\sim}{H}}{\partial t} - \frac{\partial \underset{\sim}{M}}{\partial t} + \nabla \times (\underset{\sim}{v} \times \underset{\sim}{M}) \\
\epsilon_o \nabla \cdot \underset{\sim}{E} &= \rho_e - \nabla \cdot \underset{\sim}{P} \\
\nabla \cdot (\underset{\sim}{H} + \underset{\sim}{M}) &= 0
\end{aligned}\right\} \tag{1.6}$$

together with Eqs. (1.3).

The relations between the electromagnetic vectors as observed from the laboratory frame and those in a frame moving instantaneously with the particle ("rest frame") are now

$$\left.\begin{aligned}
\underset{\sim}{E}^* &= \underset{\sim}{E} + \underset{\sim}{v} \times \mu_o \underset{\sim}{H}, & \underset{\sim}{H}^* &= \underset{\sim}{H} - \underset{\sim}{v} \times \epsilon_o \underset{\sim}{E} \\
P^* &= P, & M^* &= M
\end{aligned}\right\} \tag{1.7}$$

(*) The body is assumed to move "slowly". i.e. $v \ll c$, where v is the particle speed and c is the speed of light in vacuum.

2. General Theory and Basic Equations

Alblas [7] presents a theory of electro-magneto-thermoelasticity, based on Chu's equations, which considers a thermally and electrically conductive body with simultaneous magnetization and polarization. Magnetic dissipation is included. As in [5], the point of departure is the first law of thermodynamics in the form

$$\frac{d}{dt}\int_V \left[\rho\left(\frac{v^2}{2} + U\right) + U_e\right] dV = \int_V (\rho r + f_i v_i) dV - \int_V \left(G_i^{(e)} \dot{P}_i + G_i^{(m)} \dot{M}_i\right) dV -$$
$$-\int_V G_i^{(m)} \rho Y_i \, dV + \oint_{\partial V} \left[\tau_{ij} v_j n_i + Q^{(e)} \dot{\mathscr{P}}_i + Q_i^{(m)} \dot{\mathscr{M}}_i - Q_i n_i + X\right] d\partial V \tag{2.1}$$

and the second law in the form of the Clausius-Duhem inequality

$$\frac{d}{dt}\int_V \rho S dV \geq \int_V \frac{r}{T} \rho dV - \oint_{\partial V} \frac{Q_i n_i}{T} d\partial V \tag{2.2}$$

Here $G_i^{(e)}$ and $G_i^{(m)}$ are the effective electric and magnetic field intensity, respectively, $Q_i^{(e)}$ which, however, is put equal to zero later on, and $Q_i^{(m)}$ are the electric and magnetic surface vectors, respectively, and X is an energy supply due to the electro-magnetic field outside of V. Y_i is a vector associated with magnetic dissipation. A dot means d/dt.

After some manipulation and with some special assumptions, the balance equation (2.1) reduces to

$$\int_V \rho \dot{U} dV = \int_V \left(E_i^* - G_i^{(e)}\right) \dot{P}_i \, dV + \int_V \left(\mu_o H_i^* - G_i^{(m)}\right) \dot{M}_i \, dV +$$
$$+ \int_V \left(E_i^* j_i + \tau_{ij} v_{i,j} + \rho r - G_i^{(m)} \rho Y_i\right) dV + \tag{2.3}$$
$$+ \oint_{\partial V} \left(Q_i^{(m)} \dot{\mathscr{M}}_i - Q_i n_i\right) d\partial V$$

where the starred quantities are defined by Eqs. (1.7) and $\rho \mathscr{P}_i = P_i$, $\rho \mathscr{M}_i = M_i$.

As a basic constitutive assumption the free energy F is now introduced as

$$F = F_1 (x_{i,A}, \mathscr{P}_i, \mathscr{M}_i, \mathscr{M}_{i,j}, T) \tag{2.4}$$

If this is substituted into Eqs. (2.1) and (2.2), the following relations are deduced after a number of further special assumptions:

$$S = -\frac{\partial F_1}{\partial T}, \quad G_i^{(e)} = E_i^* - \frac{\partial F_1}{\partial \mathcal{P}_i}$$

(2.5) $$G_i^{(m)} = \mu_0 H_i^* - \frac{\partial F_1}{\partial \mathcal{M}_i} - \eta\left[\dot{\mathcal{M}}_i - (\underset{\sim}{\omega} \times \mathcal{M})_i\right] + \frac{1}{\rho} Q_{ij,j}$$

$$Q_{ij} = \rho x_{j,A} \frac{\partial F_1}{\partial M_{i,A}}, \quad \underset{\sim}{\omega} = \frac{1}{2} . \nabla \times \underset{\sim}{v}$$

where η is the "coefficient of magnetic viscosity".

To these relations are added the equation of motion

(2.6) $$\tau_{ij,i} + f_i = \rho \dot{v}_i$$

with

(2.7) $$_{ij} = \rho x_{j,A} \frac{\partial F_1}{\partial x_{i,A}} + m_{ij} + \tau_{ij}^*$$

where

(2.8) $$m_{ij} = E_i D_j - \frac{1}{2} \epsilon_0 E^2 \delta_{ij} + H_i B_j - \frac{1}{2} \mu_0 H^2 \delta_{ij} + \mu_0 (\underset{\sim}{v} \times \underset{\sim}{H})_i \mathcal{P}_j$$

represents the Maxwell stress tensor, and the equation of magnetic angular momentum

(2.9) $$\dot{M}_i = \frac{1}{2} \Gamma \left[e_{ijk} (\mathcal{M}_j G_k^{(m)} - \mathcal{M}_k G_j^{(m)}) + e_{ijk} (\mathcal{P}_j G_k^{(m)} - \mathcal{P}_k G_j^{(m)}) \right]$$

Additional constitutive assumptions for the dissipative part τ_{ij}^* of the stress and for the magnetic dissipation vector Y_i are needed together with Fourier's law for heat conduction and Ohm's law of electric current conduction.

Hutter and Pao [8] approach the problem of setting up the basic equations of magneto-thermoelasticity in a different manner. Their starting point is the Ampère formulation of Maxwell's equations for moving media(*). Furthermore,

(*) See [6], Eqs. (7.61).

the electromagnetic body force f_e, rather than being derived from the two laws of thermodynamics, is introduced directly in the Ampèrian formulation as(**)

$$\begin{aligned} \underset{\sim}{f}_e = \sigma \underset{\sim}{E} + \underset{\sim}{j} \times \underset{\sim}{B} \; + \\ + (\underset{\sim}{M}\cdot\nabla)\underset{\sim}{B} + \underset{\sim}{M} \times (\nabla \times \underset{\sim}{B}) + (\underset{\sim}{P}_{eq}\cdot\nabla)\underset{\sim}{E} - \underset{\sim}{B} \times \partial \underset{\sim}{P}_{eq}/\partial t \\ + (\underset{\sim}{P}\cdot\nabla)\underset{\sim}{E} + \underset{\sim}{v} \times (\underset{\sim}{P}\cdot\nabla)\underset{\sim}{B} - \underset{\sim}{B} \times \frac{\partial \underset{\sim}{P}}{\partial t} - \underset{\sim}{B} \times \nabla\cdot(\underset{\sim}{v}\underset{\sim}{P}) \end{aligned} \qquad (2.10)$$

The first line in Eq. (2.10) represents the Lorentz force, the second and third line are due to the magnetization and polarization models, respectively.

To Eq. (2.10) for the body force a body couple

$$\underset{\sim}{l} = \underset{\sim}{M} \times \underset{\sim}{B}' + \underset{\sim}{P} \times \underset{\sim}{E}' \qquad (2.11)$$

is adjoined together with an electromagnetic energy supply

$$\rho r_e = \underset{\sim}{j}' \cdot \underset{\sim}{E}' - \underset{\sim}{M} \cdot \dot{\underset{\sim}{B}}' + \rho \underset{\sim}{E}' \cdot \dot{\mathcal{P}} \qquad (2.12)$$

where

$$\underset{\sim}{E}' = \underset{\sim}{E} + \underset{\sim}{v} \times \underset{\sim}{B} \;, \quad \underset{\sim}{B}' = \underset{\sim}{B} - \underset{\sim}{v} \times \underset{\sim}{E}/c^2$$
$$\underset{\sim}{j}' = \underset{\sim}{j} - \rho_e \underset{\sim}{v} \;, \quad \mathcal{P} = \underset{\sim}{P}/\rho \qquad (2.13)$$

A prime indicates vectors as measured in the "rest frame" moving momentarily with the particle.

As their main constitutive equation Hutter and Pao assume the free energy density in the form

$$F = F_2 (x_{i,A}, B', T) \qquad (2.14)$$

where B' is defined by Eq. (2.13). Note that a dependence of F on the gradient of $\underset{\sim}{B}'$ and, hence, $\underset{\sim}{M}$ is not considered. "Exchange forces" are, therefore, excluded from this theory.

In the remaining part of their paper Hutter and Pao develop a linear theory of magnetoelastic interactions (a) by restricting all deformations to

---

(**) See [6]. Eq. (7.88). The electromagnetic quantities used here are different from those used in [5].

infinitesimal strains and (b) by decomposing all field variables into two parts: a rigid-body state and a perturbation state.

In a second paper [9] Pao and Hutter return to the Chu formulation of the Maxwell equations. Their results, which are, in part, less general than those of Alblas [7], are derived from a dipole – dipole model for polarization and magnetization. The constitutive assumptions are contained in the expression for the free energy which is assumed as

(2.15) $$F = F_3(x_{i,A}, \underset{\sim}{E}_e, \underset{\sim}{H}_e, T) \quad \text{for the solid}$$

and

(2.16) $$F = F_4(\rho, \underset{\sim}{E}_e, \underset{\sim}{H}_e, T) \quad \text{for the viscous fluid}$$

where

(2.17) $$\underset{\sim}{E}_e = \underset{\sim}{E} + \underset{\sim}{v} \times \mu_o \underset{\sim}{H}, \quad \underset{\sim}{H}_e = \underset{\sim}{H} - \underset{\sim}{v} \times \epsilon_o \underset{\sim}{E}$$

are the "effective" electric and magnetic field for the moving media. Equally meaningful fields could have been defined, of course, by using Eqs. (1.2.8) of [5].

In function F_4 of Eq. (2.16) the symmetric part $v_{(i,j)} = 1/2(v_{i,j} + v_{j,i})$ of the velocity gradient is first introduced as an independent variable. Later on, however, it is shown that F_4 must be independent of $v_{(i,j)}$.

Paper contains a careful derivation of jump conditions and boundary conditions.

3. Magnetothermoelastic Waves

All papers on magnetothermoelastic wave propagation are based on the linearized equations of magnetothermoelasticity. For an isotropic body these equations may be summarized as follows(*):

Equation of motion $$\rho_o \ddot{u}_i = \sigma_{ji,j} + (\underset{\sim}{j} \times \underset{\sim}{B})_i$$

(*) See [5], Section 4.1.

Hooke's law $\sigma_{ij} = 2G\epsilon_{ij} + (\lambda e_{kk} - \beta\Theta)\delta_{ij}$

$$\Theta = T - T_o \ , \quad \beta = (3\lambda + 2G)\alpha_T$$

Ohm's law $\underset{\sim}{j} = \sigma(\underset{\sim}{E} + \underset{\sim}{\dot{u}} \times \underset{\sim}{B} - \kappa \nabla\Theta)$

$$D_i = \epsilon_o E_i \ , \quad B_i = \mu H_i$$

Heat conduction $k\nabla^2\Theta = \rho_o c\,\dot{\Theta} + \beta T_o \dot{e}_{kk}$

Energy balance $\rho_o \dot{U} = \sigma_{ij}\,\dot{\epsilon}_{ji} + [E_i + (\underset{\sim}{\dot{u}} \times \underset{\sim}{B})_i\,]\cdot j_i - Q_{i,i}$

We list now several papers dealing with magnetothermoelastic waves that have appeared recently.

Based on a previous paper by Chander [10], which determines the energy loss in infinitesimal longitudinal plane waves, Puri [11] corrects and improves the results of Chander.

Nayfeh and Nemat-Nasser [12] generalize the model of infinitesimal magnetothermoelastic wave propagation by Pária and Willson(*) by adding a relaxation time of heat conduction and the electric displacement current. This changes the thermal and electromagnetic field equations from the parabolic to the hyperbolic type.

The equation of heat conduction has now the form

$$Q_i + \tau\dot{Q}_i = -\,k\,T_{,i} + \pi_o j_i \tag{3.1}$$

As been mentioned already in [5] Kaliski and Nowacki have been the first to include inertia effects in the propagation of heat in their investigation of electromagnetic-thermoelastic waves [13]. Their equations are more general than those of [12] since they include anisotropic and viscoelastic behavior of the material.

Bakshi [14] considers an infinite cylinder placed in a constant magnetic field whose direction is along the axis of the cylinder. The curved surface of the

(*) See [5], Section 4.2.

cylinder, assumed to be stress-free at all times is cooled according to $T = T_0 \exp(-\Omega t)$, $\Omega > 0$.

Again, the linearized equations (in cylindrical coordinates) are used in a perturbation procedure. A Laplace transform is applied but, as one would expect, the inverse transform can only be performed for the quasistatic case with thermal coupling assumed to be small.

Paul and Muthiyalu [15] investigate unidirectional harmonic waves in an infinite isotropic plate. The frequency equation is studied in detail and numerical results for a certain aluminum alloy are given.

Hutter an Papp [16] apply the theory developed in [8] to the wave propagation in magnetic materials. They point out several defects in the strictly linearized theory as presented in [5]. However, since they are dealing with the purely isothermal case, their paper is outside of the present survey.

4. Electrothermoelastic Waves

Based on the equations of the electrodynamically quasistatic theory(*)

$$\nabla \times \underset{\sim}{H} = 0, \qquad \nabla \times \underset{\sim}{E} = -\frac{\partial \underset{\sim}{B}}{\partial t}$$

$$\nabla \cdot \underset{\sim}{D} = 0, \qquad \nabla \cdot \underset{\sim}{B} = 0$$

$$D_i = \epsilon_0 E_i + \rho_0 \mathscr{P}_i, \; B_i = \mu_0 H_i$$

McCarthy [17] studies acceleration waves in elastic dielectrics conducting heat. Following Tiersten, a constitutive function X

$$X(x_{i,A}, E_i, \vartheta) \equiv U - TS - E_i \mathscr{P}_i \tag{4.1}$$

is introduced. From it follows

$$\left.\begin{aligned} S &= -\frac{\partial X}{\partial T}, \qquad \mathscr{P}_i = -\frac{\partial X}{\partial E_i} \\ \tau_{ij} &= \rho \frac{\partial X}{\partial x_{i,A}} x_{j,A} \end{aligned}\right\} \tag{4.2}$$

(*) See [5], Section 4.1.

We quote here only one important result of the theory. The amplitude $\underset{\sim}{a}$ and the local speed of propagation V of a homothermal(**) acceleration wave traveling in the direction $\underset{\sim}{n}$ must satisfy the propagation condition

$$(Q_{ij} - \rho V^2 \delta_{ij})\, a_j = 0 \tag{4.3}$$

where

$$Q_{ij} = \rho \frac{\partial^2 X}{\partial x_{i,A}\, \partial x_{j,B}}\, n_p n_q x_{p,A} x_{q,B} + \Gamma L_i L_j \tag{4.4}$$

and

$$\left.\begin{aligned} \Gamma &= \left(1 - \rho \frac{\partial^2 X}{\partial E_i\, \partial E_j} n_i n_j\right)^{-1} \\ L_i &= \frac{\partial \tau_{is}}{\partial E_j} n_j n_s + \rho \mathscr{P}_j n_j n_i \end{aligned}\right\} \tag{4.5}$$

$\underset{\sim}{Q}$ represents the accoustical tensor for the direction n. Any real eigenvector of $\underset{\sim}{Q}$ is a possible amplitude vector with ρV^2 as the corresponding eigenvalue. The principal directions of $\underset{\sim}{Q}$ are called the acoustical axes. Since Q is symmetric, it follows that there exists at least one set of real mutually orthogonal acoustic axes for each direction of propagation n. These axes are not, in general, paralled to the principal axes of strain, even for an isotropic body, unless the electric field vanishes ahead of the wavefront.

5. Magneto– and Electroviscoelasticity

Only a very limited number of viscoelastic materials seems to exist, which show ferromagnetic behavior (iron at elevated temperature being the most prominent one). After early attempts by Kaliski and Petykiewicz [18] a general theory, following the pattern set by Coleman, has been developed by McCarthy

---

(**) As in the magnetothermoelastic case, acceleration waves are homothermal. i.e., ϑ and $\vartheta_{,x}$ are continuous across the wave front, provided the heat conduction modulus is positive definite, cf. [5], p. 53.

[19]. It differs from the elastic case in that a constitutive functional for the free energy

$$F(t) = \underset{s=0}{\overset{\infty}{F}} \left\{ x_{i,A}(t-s),\ \mathcal{M}_i(t-s),\ \mathcal{M}_{i,j}(t-s),\ T(t-s) \right\} \qquad (5.1)$$

is introduced. The material is assumed to be magnetically saturated.

Chandrasekhariah [20] uses standard methods to discuss the propagation of magneto-thermo-viscoelastic waves in an infinite body of the linear Maxwell-Kelvin type. Since, to this authors knowledge, no magnetic material of this type exists, paper is of little physical relevance.

The same holds true for a paper by Roy [21] where again a linear viscoelastic behavior is assumed.

A problem of some engineering importance is studied by Prechtl [22] who considers the creep buckling of an axially loaded straight bar with initial imperfections. The bar is made of iron and is held at a constant elevated temperature below the Curie point. The nonlinear viscoelastic law of Norton is assumed. The influence of a transverse magnetic field on the lifetime (critical time) is considerable.

Many dielectrics, even at room temperature, exhibit a pronounced viviscoelastic behavior. Parkus [23] has given the free energy functional for a linear dielectric with memory in the form of a Fréchet series of integrals. Coefficients in the various memory functions have been assumed to be independent of temperature i.e., thermorheologically simple behavior is excluded. It would be possible, however, to include such behavior in the functional. Nevertheless, much more experimental evidence is needed before such a step may be taken with confidence. In a forthcoming paper [24] Parkus suggests a mixed form for the free energy functional.

In concluding this survey a paper by Oden and Kelley [25] should be mentioned which presents a finite element procedure for electrothermoelastic problems.

REFERENCES

[1] J.BAUCHERT: Das elastische Dielektrikum als orientiertes elastisches Kontinumm. Acta Mech. 16 (1973), 65.

[2] J. LENZ: The oriented elastic continuum as a model for the magnetoelastic body. Int. J. Solids Structures 8 (1972), 1235.

[3] L. KNOPOFF: The interaction between elastic wave motions and a magnetic field in electrical conductors. J. Geophys. Res. 60 (1955), 441.

[4] H. PARKUS: Magneto– und Elektroelastizität. Z. ang. Math. Mech. 53 (1973), T 18.

[5] H. PARKUS: Magneto-Thermoelasticity. CISM Courses and Lectures No. 118. Springer-Verlag Wien-New York 1972.

[6] P. PENFIELD, Jr., and H.A. HAUS: Electrodynamics of Moving Media. M.I.T. Press, Cambridge, Mass. 1967.

[7] J.B. ALBLAS: Electro-Magneto-Elasticity. In: Topics in Applied Continuum Mechanics (J.L. Zeman and F. Ziegler, editors) Springer-Verlag Wien--NewYork, 1974.

[8] K. HUTTER and Y.H. PAO: A dynamic theory for magnetizable elastic solids with thermal and electrical conduction. J. Elasticity (in press).

[9] Y.H. PAO and K. HUTTER: Electrodynamics for moving elastic solids and viscous fluids. I.E.E.E.J. (in press).

[10] S. CHANDER: Phase velocity and energy loss in magneto-thermo-elastic plane waves. Int. J. Engng Sci. 6 (1968), 409.

[11] P. PURI: Plane waves in thermoelasticity and magnetothermoelasticity. Int. J. Engng Sci 10 (1972), 467.

[12] A.H. NAYFEH and S. NEMAT-NASSER: Electromagneto-Thermoelastic plane waves in solids with thermal relaxation. J. Appl. Mech. 39 (1972), 108.

[13] S. KALISKI and W. NOWACKI: Thermal excitations in coupled fields. In: Progress in Thermoelasticity (Editor W.K. Nowacki), Warsaw (1969).

[14] S.K. BAKSHI: On the effect of magneto-thermo-elastic interactions on the cooling process of an infinite circular cylinder. Proc. Camb. Phil. Soc. 67 (1970), 67.

[15] H.S. PAUL and N. MUTHIYALU: Free vibrations of an infinite isotropic magneto-thermo- elastic plate. Acta Mech. 14 (1972), 147.

[16] K. HUTTER and L. PAPP: Wave propagation and attenuation in paramagnetic and soft ferromagnetic materials. (In press).

[17] M.F.McCARTHY: Thermodynamic influences on the propagation of waves in electroelastic materials. Int. Engng Sci. 11 (1973), 1301.

[18] S. KALISKI and J. PETYKIEWICZ: Equations of motion coupled with the field of temperature in the magnetic field involving mechanical and electro-magneto relaxation for anisotropic bodies. Proc. Vibr. Probl. 4 (1960), 3.

[19] M.F. McCARTHY: Thermodynamics of deformable magnetic materials with memory. Int. J. Engng Sci. 12 (1974), 45.

[20] D.S. CHANDRASEKHARIAH : The propagation of magneto-thermo-visco-elastic plane waves in a parallel union of the Kelvin and Maxwell bodies. Proc. Camb. Phil. Soc. 70 (1971), 343.

[21] P. ROY: Disturbances in a perfectly conducting viscoelastic plate in a magnetic field. Bull. Acad. Polon. Sci., série sci. techn. 17 (1969), 233 – [341].

[22] A. PRECHTL: Creep buckling of plates in a transverse magnetic field. Sitzungsberichte Oesterr. Akademie der Wissensch. Math. –Naturw. Klasse

(in press).

[23] H. PARKUS: Constitutive equations of the linear viscoelastic dielectric. In: Trends in Elasticity and Thermoelasticity. Wolters-Noordhoof Publishing 1971, Groningen.

[24] H. PARKUS: Constitutive equations for a thermorheologically simple dielectric. L. Sobrero Anniversary Volume. Udine (to appear).

[25] J.T. ODEN and B.E. KELLEY: Finite element formulation of general electrothermoelasticity problems. Int. J. Numerical Meth. Engng 3 (1971), 161.

NOTATION

All equations are written in the international MSK system (Giorgi system). Basic units are meter m, second s, kilogram kg and Ampère A. (Volt V = m^2 kg/As^3 , Newton N = kg m/s^2).

A	surface area m^2
$\underset{\sim}{B}$	magnetic induction, Vs/m^2
$\underset{\sim}{D}$	electric displacement, As/m^2
$\underset{\sim}{E}$	electric field strength, V/m
E	modulus of elasticity, N/m^2
F	free energy per unit of mass, Nm/kg
$\underset{\sim}{F}$	force, N
G	shear modulus, N/m^2
$\underset{\sim}{H}$	magnetic field strength, A/m
$\underset{\sim}{J}$	surface current density, A/m
$\underset{\sim}{L}$	moment, Nm
$\underset{\sim}{M}$	magnetization per unit volume, A/m
$\underset{\sim}{\mathcal{M}}$	magnetization per unit of mass, $\mathcal{M} = M/\rho$
$\underset{\sim}{P}$	polarization per unit of volume, As/m^2
$\underset{\sim}{\mathcal{P}}$	polarization per unit of mass, $\mathcal{P} = P/\rho$
Q	heat flux, N/ms
S	entropy per unit of mass, N/ms°K
T	absolute temperature, °K, T_0 reference temperature
U	internal energy per unit of mass, Nm/kg
U_e	electromagnetic energy per unit of volume, N/m^2
V	volume, m^3
V	wave speed, m/s
X_A	spatial coordinate, m,
a_{ij}	exchange tensor, N/A
c	speed of light in vacuo, m/s
$\underset{\sim}{f}$	volume force density, N/m^3
$\underset{\sim}{f}_L$	Lorentz force density, N/m^3
$\underset{\sim}{j}$	electric current density, A/m^2
k	thermal conductivity, N/s°K
m	mass, kg

$\underset{\sim}{n}$	unit normal vector,
r	strength of heat source distribution, Nm/kg s
t	time, s
t_{ij}	stress tensor, N/m^2
u_i	displacement vector, m
v_i	particle velocity,
x_i	material coordinate,
$x_{i,A}$	deformation gradient
α_T	coefficient of thermal expansion, $1/°K$
γ	wave number, $1/m$
ϵ	dielectric constant, As/Vm
ϵ_o	dielectric constant of free space, $\epsilon_o = 8.859 \times 10^{-12}$
ϵ_{ij}	strain tensor
ϑ	temperature °K, $\vartheta = T - T_o$
κ	coefficient in Ohm's law, V/m°K
μ	permeability, Vs/Am
μ_o	permeability of free space, $\mu_o = 1.257 \times 10^{-6}$, $\epsilon_o \mu_o = 1/c^2$
ν	Poisson's ratio
$\underset{\sim}{\pi}$	measure of polarization
ρ	mass density, kg/m^3
ρ_e	charge density, As/m^3
σ	electrical conductivity, A/Vm
σ_{ij}	stress tensor, N/m^2
τ_{ij}	Cauchy stress tensor, N/m^2
ω	circular frequency, $1/s$
ω_{ij}	rotation tensor.

ON THE THERMOELASTICITY OF NON–LINEAR DISCRETE AND CONTINUOUS CONSTRAINED SYSTEMS

Cz. WOZNIAK
(Warszawa)

Abstract. In the note equations of thermoelasticity of discrete and continuous systems are derived with the aid of the concept of ideal constraints imposed on mechanical as well as non-mechanical quantities. The thermoelastic systems are obtained as models of a certain material system M, in which a set of non-intersecting finite elements is distinguished. We assume that any typical element of M undergoes only homogeneous deformations and linear distributions of temperature; the concept of ideal constraints is applied to derive basic equations of thermoelasticity for this element. The constraints are also introduced to describe the interactions among elements: the stress vector and the heat flux on the corresponding boundaries of interacting elements are assumed to be continuous. The equations of the discrete thermoelastic system are obtained under condition that the pairs of interacting elements form a lattice in R^N, $0 < N \leq 3$. The N-th dimensional thermoelastic continuum is derived from the suitable discrete system, provided that elements of M are sufficiently small and that certain regularity conditions hold.

1. Thermoelasticity of uniform deformations and temperature gradients

An object of our investigations will be certain material system M, which is to be analysed only by means of some discrete and continuous models. To obtain a discrete model of M we shall take into consideration a finite set of disjoined material elements $\mathcal{P}(Y) \subset M$, $Y \in D_R$, D_R being the finite set of points in R^N, $0 < N \leq 3$. Each $Y \in D_R$ will be identified with the position vector of the centre of mass of an element $\mathcal{P}(Y)$ in the fixed reference configuration of M. We assume that each $\mathcal{P}(Y)$ can be treated, with a sufficient accuracy, as a homogeneous hyperlastic material continuum with the uniform distribution of deformations and the linear distribution of the temperature. We also assume that the physical properties of the part $M - U\mathcal{P}(Y)$ of M can be neglected when we deal with models of the material system M. It follows that the interactions between any pair of elements (only binary interactions will be taken into account) we obtain by assuming that the suitable stress vector and heat flux can be treated as continuous between the elements.

In this Section we shall confine ourselves to the one typical element $\mathcal{P}(Y)$, in Sec. 2 we arrive at the discrete models of M by taking into account the interactions among elements, and in Sec. 3 we are to anayse the continuous models of M, which will be obtained form the suitable discrete systems, provided that certain regularity conditions hold.

Let x^k, t be inertial coordinates in the space-time(*), $\{x^k\}$ being the Carthesian orthogonal coordinates in the physical space. Let $M \supset \mathscr{P}(Y) \to \mathscr{P}_R \subset R^3$, $\mathscr{P}_R$ being the region in R^3., be a fixed reference configuration of any element $\mathscr{P}(Y)$. The deformation function ζ_k of this element is given by

$$\zeta_k = F_{k\alpha}(Y,t)Z^\alpha + X_k(Y,t); \ \det F > 0; \ Z \equiv (Z^\alpha) \in \bar{\mathscr{P}}_R, Y \equiv (Y^K) \in D_R, t \in R, \tag{1.1}$$

and the distribution of the absolute temperetare ϑ is equal to

$$\vartheta = \theta_\alpha(Y,t)Z^\alpha + \theta(Y,t); \ \vartheta > 0; \quad Z \in \bar{\mathscr{P}}_R, \ Y \in D_R, \ t \in R. \tag{1.2}$$

The thermomechanics of the aterial element $\mathscr{P}(Y)$ let be governed by a well known system of the physical laws expressed in the local form

$$\begin{aligned} &\mu_R \ddot{\zeta}^k = T_{R\,,\alpha}^{k\alpha} + \mu_R b^k, \quad T_R^{[k\alpha}F^{\ell]}{}_\alpha = 0; \\ &\mu_R \dot{\epsilon} = h_{R,\alpha}^\alpha + \mu_R q + T_R^{k\alpha}\dot{F}_{k\alpha}, \\ &\mu_R \vartheta\dot{\eta} - h_{R,\alpha}^\alpha - \mu_R q \geqslant 0, \ h_R^\alpha \theta_\alpha \geqslant 0; \quad Z \in \mathscr{P}_R, \ t \in R, \ Y \in D_R; \\ &T_R^{k\alpha} n_{R\alpha} = p_R^k, \ h_R^\alpha n_{R\alpha} = h_R; \ Z \in \partial\mathscr{P}_R, \ t \in R, \ Y \in D_R, \end{aligned} \tag{1.3}$$

where $\mu_R = \mu_R(Y;Z)$ is a mass density, $T_R = T_R(Y;Z,t)$ is a first Piola-Kirchhoff stress tensor, $b^k = b^k(Y;Z,t)$ are body forces, $\epsilon = \epsilon(Y;Z,t)$ and $\eta = \eta(Y;Z,t)$ are a specific internal energy and a specific entropy, respectively, $h_R^\alpha = h_R^\alpha(Y;Z,t)$ and $q = q(Y;Z,t)$ are a heat flux and a heat absorption, respectively, $p_R^k = p_R^k(Y;Z,t)$ and $h_R = h_R(Y;Z,t)$ are surface tractions and boundary heat absorption, respectively; the subscript "R" inform us, that the preceding quantity is related to the reference configuration of $\mathscr{P}(Y)$.

Because Eqs. (1.1), (1.2) represent the constraints imposed on the fields $\zeta_k(Y;Z,t)$, $\vartheta(Y;Z,t)$ in each $\mathscr{P}(Y)$, we have to postulate the principles of reaction

(*) Indices α, β, k run over $\{1,2,3\}$, and indices K, L run over $\{1,\dots,N\}$ the integer N, $0 < N \leqslant 3$, being fixed in each problem. Summation convention holds only for tensorial indices.

due to those constraints,

$$\begin{aligned} b^k &= \overset{\circ}{b}{}^k + r^k \ , \quad q = \overset{\circ}{q} + v \qquad \text{in} \quad \mathscr{P}_R \ , \\ p_R^k &= \overset{\circ}{p}{}_R^k + s_R^k \ , \quad h_R = \overset{\circ}{h}_R + w_R \qquad \text{on} \quad \partial\mathscr{P}_R \ , \end{aligned} \tag{1.4}$$

in each element $\mathscr{P}(Y)$, where r, s_R, v, w_R are the reactions.

The constraints for deformations and a temperature will be called ideal if the relations

$$\oint_{\partial\mathscr{P}_R} s_R^k \delta\zeta_k d\sigma_R + \int_{\mathscr{P}_R} \mu_R r^k \delta\zeta_k dv_R = 0 \ , \qquad \oint_{\partial\mathscr{P}_R} w_R \delta\vartheta d\sigma_R + \int_{\mathscr{P}_R} \mu_R v \delta\vartheta dv_R = 0 \ , \tag{1.5}$$

hold for any variations $\delta\zeta_k$, $\delta\vartheta$ consistent with constraints for deformations and a temperature, respectively. We deal here with a generalization of the known concept of ideal constraints, cf. [3], on non-mechanical fields. In what follows all models of the material system M and its elements will be obtained under assumption that the constraints at each element $\mathscr{P}(Y)$ are perfect, [2], and we introduce the specific free energy function $\varphi(Y;Z,t) = \tilde{\varphi}(Y;Z,F,\vartheta)$. The constitutive equations are

$$\eta = -\frac{\partial\tilde{\varphi}}{\partial\vartheta} = \tilde{\eta}(Y;Z,F,\vartheta) \ , \quad T_R^{k\alpha} = \mu_R \frac{\partial\tilde{\varphi}}{\partial F_{k\alpha}} \ , \quad h_R^{\alpha} = \tilde{h}_R^{\alpha}(Y;Z,\theta_{\alpha},\theta,F) \ , \tag{1.6}$$

where the response function $\tilde{h}_R^{\alpha}$ has to satisfy the condition $\tilde{h}_R^{\alpha}\theta_{\alpha} > 0$ for any $\theta_{,\alpha}$.

Using (1.1) – (1.6) we arrive at the following equations of motion for an element $\mathscr{P}(Y)$:

$$\begin{aligned} \rho_R \ddot{X}^k &= \overset{\circ}{f}{}_R^k \ , \\ \ell^2 I_R^{\alpha\beta} \ddot{F}^k{}_{\beta} &= \overset{\circ}{S}{}_R^{k\alpha} - \rho_R \frac{\partial\sigma}{\partial F_{k\alpha}} \ , \end{aligned} \tag{1.7}$$

where we have introduced the following denotations

$$\rho_R = \rho_R(Y) \equiv \frac{1}{\ell^3} \int_{\mathscr{P}_R} \mu_R dv_R \ , \quad \ell^2 I_R^{\alpha\beta} = \ell^2 I_R^{\alpha\beta}(Y) \equiv \frac{1}{\ell^3} \int_{\mathscr{P}_R} Z^{\alpha} Z^{\beta} \mu_R dv_R \ ,$$

$$\rho_R \sigma(Y,F,\theta,\theta_{\alpha}) \equiv \frac{1}{\ell^3} \int_{\mathscr{P}_R} \tilde{\varphi}\mu_R dv_R \ ,$$

$$\overset{\circ k}{f}_R = \overset{\circ k}{f}_R(Y,t) \equiv \frac{1}{\ell^3}\left(\oint_{\partial\mathscr{P}_R} \overset{\circ k}{p}_R d\sigma_R + \int_{\mathscr{P}_R} \mu_R \overset{\circ k}{b} dv_R\right),$$

(1.8)

$$\overset{\circ k\alpha}{S}_R = \overset{\circ k\alpha}{S}_R(Y,t) \equiv \frac{1}{\ell^3}\left(\oint_{\partial\mathscr{P}_R} Z^\alpha \overset{\circ k}{p}_R d\sigma_R + \int_{\mathscr{P}_R} \mu_R Z^\alpha \overset{\circ k}{b} dv_R\right),$$

and ℓ^3 is a volume of the region $\mathscr{P}_R$(*).We also obtain the entropy balance equations

(1.9)

$$\overset{\circ}{\gamma}_R + \alpha_R\dot{\theta} + B_R^{k\alpha}\dot{F}_{k\alpha} + \ell a_R^\alpha\theta_\alpha = 0,$$

$$\overset{\circ}{q}{}^\alpha_R - j^\alpha_R + \ell a^\alpha_R\dot{\theta} + \ell G_R^{\alpha k\beta}\dot{F}_{k\beta} + \ell^2 A_R^{\alpha\beta}\dot{\theta}_\beta = 0,$$

where

$$\alpha_R \equiv \frac{1}{\ell^3}\int_{\mathscr{P}_R} \vartheta\frac{\partial^2\varphi}{\partial\theta^2}\mu_R dv_R,\quad \ell a_R^\alpha \equiv \frac{1}{\ell^3}\int_{\mathscr{P}_R} \vartheta\frac{\partial^2\varphi}{\partial\theta^2}Z^\alpha\mu_R dv_R,\quad \ell^2 A_R^{\alpha\beta} \equiv \frac{1}{\ell^3}\int_{\mathscr{P}_R} \vartheta\frac{\partial^2\tilde{\varphi}}{\partial\theta^2}Z^\alpha Z^\beta\mu_R dv_R,$$

$$B_R^{k\alpha} \equiv \frac{1}{\ell^3}\int_{\mathscr{P}_R} \vartheta\frac{\partial^2\tilde{\varphi}}{\partial\theta\partial F_{k\alpha}}\mu_R dv_R,\quad \ell G_R^{\alpha k\beta} \equiv \frac{1}{\ell^3}\int_{\mathscr{P}_R} \vartheta\frac{\partial^2\tilde{\varphi}}{\partial\theta\partial F_{k\beta}}Z^\alpha\mu_R dv_R,$$

(1.10)

$$\overset{\circ}{\gamma}_R \equiv \frac{1}{\ell^3}\left(\oint_{\partial\mathscr{P}_R} \overset{\circ}{h}_R d\sigma_R + \int_{\mathscr{P}_R} \overset{\circ}{q}_R\mu_R dv_R\right),\quad \overset{\circ}{q}{}^\alpha_R \equiv \frac{1}{\ell^3}\left(\oint_{\partial\mathscr{P}_R} Z^\alpha\overset{\circ}{h}_R d\sigma_R + \int_{\mathscr{P}_R} Z^\alpha\overset{\circ}{q}\mu_R dv_R\right),$$

$$j^\alpha_R \equiv \frac{1}{\ell^3}\int_{\mathscr{P}_R} \tilde{h}^\alpha_R dv_R.$$

The lenght parameter ℓ enables us to treat all quantities in Eqs. (1.7), (1.9) as a mean densities related to the reference configuration; Eqs. (1.7) are valid under condition that the point $Z = 0$ is a mass center of $\mathscr{P}(Y)$ in the reference configuration.

The material element $\mathscr{P}(Y)$ is subjected to known external agents (such as surface or body loads) and to interactions with other elements, which are not known a priori. In particular we have $\overset{\circ k}{p}_R = \check{p}^k_R + \tilde{p}^k_R$, $\overset{\circ}{h}_R = \check{h}_R + \tilde{h}_R$, where $\check{p}^k_R$, $\check{h}_R$ characterize the unknown interactions of $\mathscr{P}(Y)$ with other elements and $\tilde{p}^k_R$, $\tilde{h}_R$ are known surface loads and input of heat, respectively. Assuming that the fields $\overset{\circ}{b}$, $\overset{\circ}{q}_R$

(*) Mind, that σ is not a strain energy function, being dependent on the temperature gradient θ_α.

represent only known external agents, we obtain

$$\overset{\circ}{f}{}^{k}_{R} = \overset{\vee}{f}{}^{k}_{R} + f^{k}_{R},\ \overset{\circ}{\gamma}_{R} = \overset{\vee}{\gamma}_{R} + \gamma_{R},\ \overset{\circ}{S}{}^{k\alpha}_{R} = \overset{\vee}{S}{}^{k\alpha}_{R} + S^{k\alpha}_{R},\ \overset{\circ}{q}{}^{\alpha}_{R} = \overset{\vee}{q}{}^{\alpha}_{R} + q^{\alpha}_{R}, \tag{1.11}$$

where

$$\overset{\vee}{f}{}^{k}_{R} \equiv \frac{1}{\varrho^3}\oint_{\partial\mathscr{P}_R} \overset{\vee}{p}{}^{k}_{R}\, d\sigma_R , \quad f^{k}_{R} \equiv \frac{1}{\varrho^3}\Big(\oint_{\partial\mathscr{P}_R} \tilde{p}^{k}_{R} d\sigma_R + \int_{\mathscr{P}_R} \overset{\circ}{b}{}^{k}\mu_R dv_R\Big);$$

$$\overset{\vee}{\gamma}_{R} \equiv \frac{1}{\varrho^3}\oint_{\partial\mathscr{P}_R} \overset{\vee}{h}_{R} d\sigma_R , \quad \gamma_{R} \equiv \frac{1}{\varrho^3}\Big(\oint_{\partial\mathscr{P}_R} \tilde{h}_{R} d\sigma_R + \int_{\mathscr{P}_R} \overset{\circ}{q}\mu_R dv_R\Big) ; \tag{1.12}$$

$$\overset{\vee}{S}{}^{k\alpha}_{R} \equiv \frac{1}{\varrho^3}\oint_{\partial\mathscr{P}_R} Z^{\alpha}\overset{\vee}{p}{}^{k}_{R} d\sigma_R, \quad S^{k\alpha}_{R} \equiv \frac{1}{\varrho^3}\Big(\oint_{\partial\mathscr{P}_R} Z^{\alpha}\tilde{p}^{k}_{R} d\sigma_R + \int_{\mathscr{P}_R} Z^{\alpha}\overset{\circ}{b}{}^{k}\mu_R dv_R\Big) ;$$

$$\overset{\vee}{q}{}^{\alpha}_{R} \equiv \frac{1}{\varrho^3}\oint_{\partial\mathscr{P}_R} Z^{\alpha}\overset{\vee}{h}_{R} d\sigma_R, \quad q^{\alpha}_{R} \equiv \frac{1}{\varrho^3}\Big(\oint_{\partial\mathscr{P}_R} Z^{\alpha}\tilde{h}_{R} d\sigma_R + \int_{\mathscr{P}_R} Z^{\alpha}\overset{\circ}{q}\mu_R dv_R\Big) .$$

The functions introduced above are defined on $D_R^R \times R$; the first terms in (1.11) are due only to interactions among elements and the second ones characterize all other agents acting at $\mathscr{P}(Y)$.

Equations (1.7), (1.9), after taking into account Eqs. (1.11), constitute, for each $Y \in D_R$, the system of 16 equations for 32 unknowns X_k, $F_{k\alpha}$, θ, θ_α, $\overset{\vee}{f}{}^{k}_{R}$, $\overset{\vee}{S}{}^{k\alpha}_{R}$, $\overset{\vee}{\gamma}_{R}$, $\overset{\vee}{q}{}^{\alpha}_{R}$. The missing equations will be obtained in the next Section by postulating the character of interaction among elements $\mathscr{P}(Y)$, $Y \in D_R$.

2. Discrete Thermoelastic Systems

We shall deal only with binary interactions between elements $\mathscr{P}(Y)$, $Y \in D_R$, assuming that they are distributed over the set of elements $\mathscr{P}(Y)$ in some regular manner. To describe this distribution we assign to each $Y \in D_R$ a system of 2N different vectors which constitute two vectors basis $\Delta_K Y$ and $\bar{\Delta}_K Y$ in R^N, $K = 1, \ldots, N$. We assume that the elements $\mathscr{P}(Y)$, $\mathscr{P}(\bar{Y})$ interact only if $\bar{Y} = Y + \Delta_K Y$ or $Y = \bar{Y} + \bar{\Delta}_K \bar{Y}$; moreover, for any two interacting elements $\mathscr{P}(Y)$, $\mathscr{P}(\bar{Y})$, we have $(\bar{Y} = Y + \Delta_K Y) \Leftrightarrow (Y = \bar{Y} + \bar{\Delta}_K \bar{Y})$. If the relation $\Delta_K Y = -\bar{\Delta}_K Y$ holds for each Y, then D_R is a regular lattice of points in a certain region of R^N, $0 < N \leqslant 3$.

Let $u : D_R \to R$, $u^K : D_R \to R$, be arbitrary real valued functions, and let $D_R^K : \{Y; Y \in D_R \wedge Y + \Delta_K Y \in D_R\}$, $\bar{D}_R^K : \{Y; Y \in D_R \wedge Y + \bar{\Delta}_K Y \in D_R\}$.

We shall use the following operators

$$\Delta_K u(Y) \equiv \frac{1}{\ell}[u(Y+\Delta_K Y) - u(Y)], \nabla^{\alpha}_K u^K(Y) \equiv \frac{1}{2}\delta^{\alpha}_K[u^K(Y) + u^K(Y+\Delta_K Y)]; \quad Y \in D^K_R ,$$

(2.1)

$$\bar{\Delta}_K u(Y) \equiv \frac{1}{\ell}[u(Y+\bar{\Delta}_K Y) - u(Y)], \bar{\nabla}^{\alpha}_K u^K(Y) \equiv \frac{1}{2}\delta^{\alpha}_K[u^K(Y) + u^K(Y+\bar{\Delta}_K Y)]; \quad Y \in \bar{D}^K_R .$$

We shall also denote by S_K and $\bar{S}_K$ the parts of the surface $\partial\mathscr{P}_R$, and assume that they are regions on the parametric planes $Z^K = \ell/2$ and $Z^K = -\ell/2$, respectively. Interaction of an element $\mathscr{P}(Y)$ with an element $\mathscr{P}(Y + \Delta_K Y)$ is assumed to be realized exclusively across the surface S_K; analogously, interaction of an element $\mathscr{P}(Y)$ with an element $\mathscr{P}(Y + \bar{\Delta}_K Y)$ is realized only across $\bar{S}_K$. Let the densities of interaction on S_K be denoted by $\check{p}^k_R = p^{kK}_R$, $\check{h}_R = g^K_R$, and on $\bar{S}_K$ be denoted by $\check{p}^k_R = \bar{p}^{kK}_R$, $\check{h}_R = \bar{g}^K_R$, cf. (1.12).

Using Eqs. (1.12) we obtain

$$\check{f}^k_R = \sum_{K=1}^{N} \frac{1}{\ell}\left(T^{kK}_R - \bar{T}^{kK}_R\right), \quad \check{S}^{kK}_R = \frac{1}{2}\left(T^{kK}_R + \bar{T}^{kK}_R\right),$$

(2.2)

$$\check{\gamma}_R = \sum_{K=1}^{N} \frac{1}{\ell}\left(h^K_R - \bar{h}^K_R\right), \quad \check{q}^K_R = \frac{1}{2}\left(h^K_R + \bar{h}^K_R\right); \; Y \in D_R ,$$

where

$$T^{kK}_R \equiv \frac{1}{\ell^2}\int_{S_K} p^{kK}_R d\sigma_R, \quad \bar{T}^{kK}_R \equiv -\frac{1}{\ell^2}\int_{\bar{S}_K} \bar{p}^{kK}_R d\sigma_R, \quad h^K_R \equiv \frac{1}{\ell^2}\int_{S_K} g^K_R d\sigma_R,$$

(2.3)

$$\bar{h}^K_R \equiv -\frac{1}{\ell^2}\int_{\bar{S}_K} \bar{g}^K_R d\sigma_R,$$

and where we have assumed that

$$\int_{S_K} Z^{\beta} p^{kK}_R d\sigma_R = \int_{\bar{S}_K} Z^{\beta} \bar{p}^{kK}_R d\sigma_R = \int_{S_K} Z^{\beta} g^K_R d\sigma_R = \int_{\bar{S}_K} Z^{\beta} \bar{g}^K_R d\sigma_R = 0$$

for each $\beta \neq K$. If $Y \sim\in D^K_R$ then $T^{kK}_R = h^K_R \equiv 0$; if $Y \sim\in \bar{D}^K_R$ then $\bar{T}^{kK}_R = \bar{h}^K_R \equiv 0$. It must be stressed that the symbols T^{kK}_R , h^K_R here and in what follows have the different meaning than the symbols $T^{k\alpha}_R$, h^{α}_R in Sec. 1. Using the denotations (2.3), the interaction between elements $\mathscr{P}(Y)$, $\mathscr{P}(Y + \Delta_K Y)$ and between elements $\mathscr{P}(Y)$,$\mathscr{P}(Y + \bar{\Delta}_K Y)$, will be postulated in the form

$$T^{kK}_R(Y,t) = \bar{T}^{kK}_R(Y + \Delta_K Y, t) , \; h^K_R(Y,t) = \bar{h}^K_R(Y + \Delta_K Y, t) ; \; Y \in D^K_R ,$$

$$\bar{T}^{kK}_R(Y,t) = T^{kK}_R(Y + \bar{\Delta}_K Y, t)\ ,\ \bar{h}^K_R(Y,t) = h^K_R(Y + \bar{\Delta}_K Y, t)\ ;\ Y \in \bar{D}^K_R\ , \tag{2.4}$$

(2.4)

respectively (cf. Sec. 1). Using (2.4), from Eqs. (2.2) we obtain

$$\begin{aligned} \check{f}^k_R &= \Delta_K \bar{T}^{kK}_R = \bar{\Delta}_K T^{kK}_R\ , & \check{S}^{k\alpha}_R &= \nabla^\alpha_K \bar{T}^{kK}_R = \bar{\nabla}^\alpha_K T^{kK}_R\ ,\\ \check{\gamma}_R &= \Delta_K \bar{h}^K_R = \bar{\Delta}_K h^K_R\ , & \check{q}^\alpha_R &= \nabla^\alpha_K \bar{h}^K_R = \bar{\nabla}^\alpha_K h^K_R\ . \end{aligned} \tag{2.5}$$

(2.5)

Equations (2.5) represent the constraints for the fields $\check{f}^k_R$, $\check{S}^{k\alpha}_R$, $\check{\gamma}_R$ and $\check{q}^\alpha_R$. Generalizing an approach given in [3], we shall call such constraints ideal if the relations

$$\begin{aligned} &\sum_{Y \in D_R} [X_k(Y,t)\delta\check{f}^k_R(Y,t) + F_{k\alpha}(Y,t)\delta\check{S}^{k\alpha}_R(Y,t)] = 0,\\ &\sum_{Y \in D_R} [\theta(Y,t)\delta\check{\gamma}_R(Y,t) + \theta_\alpha(Y,t)\delta\check{q}^\alpha_R(Y,t)] = 0, \end{aligned} \tag{2.6}$$

(2.6)

hold for any $\delta\check{f}^k_R$, $\delta\check{S}^{k\alpha}_R$, $\delta\check{\gamma}_R$, $\delta\check{q}^\alpha_R$ consistent with the constraints (2.5). From (2.6) and (2.5) we obtain

$$\begin{aligned} \Delta_K X_k &- \nabla^\alpha_K F_{k\alpha} = I_{kK}\ ,\\ \Delta_K \theta &- \nabla^\alpha_K \theta_\alpha = i_K\ , \end{aligned} \tag{2.7}$$

(2.7)

where

$$\sum_Y I_{kK}(Y,t)\delta T^{kK}_R(Y,t) = 0,\ \sum_Y i_K(Y,t)\delta h^K_R(Y,t) = 0 \tag{2.8}$$

(2.8)

The fields $I_{kK}(Y,t)$, $i_K(Y,t)$ will be called the discrete incompatibilities of deformations and temperature gradients. In many cases we introduce an extra restriction on the fields $T^{kK}_R(Y,t)$, $h^K_R(Y,t)$; this restriction can be given by

$$\alpha^\mu(Y, T^{kK}_R(Y,t)) = 0\ ,\ \mu = 1,\ldots,m;\ \beta^\rho(Y, h^K_R(Y,t)) = 0,\ \rho = 1,\ldots,r, \tag{2.9}$$

(2.9)

where α^μ and β^ρ are known differentiable functions, and δT^{kK}_R , δh^K_R in the relations (2.8) have to be consistent with (2.9). If there are no extra constraints of the form (2.9), then from (2.8) we obtain $I_{kK}(Y,t) = 0, i_K(Y,t) = 0$; in this case Eqs. (2.7) will be called the equations of compatibility of the discrete thermoelastic system. Substituting the right-hand sides of Eqs. (2.5) into (1.7) (taking into account (1.11)), we obtain the equations of motion

(2.10)
$$\bar{\Delta}_K T_R^{kK} + f_R^k = \rho_R \ddot{X}^k ,$$
$$\ell^2 I_R^{\alpha\beta} \ddot{F}^k{}_\beta = \bar{\nabla}_K^\alpha T_R^{kK} - \rho_R \frac{\partial\sigma}{\partial F_{k\alpha}} + S_R^{k\alpha} ,$$

and the heat conduction equations (the entropy balance equations)

(2.11)
$$\bar{\Delta}_K h_R^K + \gamma_R + \alpha_R \dot{\theta} + B_R^{k\beta} \dot{F}_{k\beta} + \ell a_R^\alpha \dot{\theta}_\alpha = 0 ,$$
$$\bar{\nabla}_K^\alpha h_R^K - j_R^\alpha + q_R^\alpha + \ell a_R^\alpha \dot{\theta} + \ell^2 A_R^{\alpha\beta} \dot{\theta}_\beta + \ell B_R^{\alpha k\beta} \dot{F}_{k\beta} = 0$$

for a discrete thermoelastic continuum.

In some problems we can also deal with the constraints imposed on the fields $X(Y,t)$, $F(Y,t)$, $\theta(Y,t)$, $\theta_\alpha(Y,t)$. Let us confine ourselves to the constraints of the form

(2.12)
$$\alpha_\nu (Y, X_k(Y,t) , F_{k\alpha}(Y,t)) = 0, \; \nu = 1,\ldots,n ;$$
$$\beta_\pi (Y, \theta(Y,t) , \theta_\alpha (Y,t)) = 0, \; \pi = 1,\ldots,\pi ,$$

where α_ν , β_π are known differentiable functions. Postulating the principles of reaction for constraints (2.12)

(2.13)
$$f_R^k = \bar{f}_R^k + \hat{f}_R^k , \quad S_R^{k\alpha} = \bar{S}_R^{k\alpha} + \hat{S}_R^{k\alpha} ,$$
$$\gamma_R = \bar{\gamma}_R + \hat{\gamma}_R , \quad q_R^\alpha = \bar{q}_R^\alpha + \hat{q}_R^\alpha ,$$

where $\bar{f}_R^k$, $\bar{S}_R^{k\alpha}$, $\bar{\gamma}_R$, $\bar{q}_R^\alpha$ are known external "loads" and $\hat{f}_R^k$, $\hat{S}_R^{k\alpha}$, $\hat{\gamma}_R$, $\hat{q}_R^\alpha$, are unknown reactions due to the constraints, and assuming that the constraints are ideal

(2.14)
$$\sum_Y [\hat{f}_R^k(Y,t)\delta X_k(Y,t) + \hat{S}_R^{k\alpha}(Y,t)\delta F_{k\alpha}(Y,t)] = 0 ,$$
$$\sum_Y [\hat{\gamma}_R(Y,t)\delta\theta(Y,t) + \hat{q}_R^\alpha(Y,t)\delta\theta_\alpha(Y,t)] = 0 ,$$

(δX_k , $\delta F_{k\alpha}$, $\delta\theta$, $\delta\theta_\alpha$ are variations consistent with (2.12)), we arrive at the relations

$$f_R^k = \bar{f}_R^k + \lambda^\nu \frac{\partial a_\nu}{\partial X_k} , \quad S_R^{k\alpha} = \bar{S}_R^{k\alpha} + \lambda^\nu \frac{\partial a_\nu}{\partial F_{k\alpha}} ,$$

$$\gamma_R = \bar{\gamma}_R + \mu^\pi \frac{\partial \beta_\pi}{\partial \theta} , \quad q_R^\alpha = \bar{q}_R^\alpha + \mu^\pi \frac{\partial \beta_\pi}{\partial \theta_\alpha} , \tag{2.15}$$

where $\lambda^\nu = \lambda^\nu$ (Y,t), $\mu^\pi = \mu^\pi$ (Y,t) are unknown Lagrange's multipliers. The right-hand sides of (2.15) have to be substituted into Eqs. (2.10), (2.11), and Eqs. (2.12) have to be included into a basic system of equations of the discrete thermoelasticity.

It must be stressed that the interactions among elements, postulated here in the form (2.4), can be also realised on two other ways. Firstly, the fields of deformation and temperature can be assumed continuous between interacting elements. Secondly, we can introduce the long distance interactions between $\mathscr{P}$(Y) and $\mathscr{P}(Y + \Delta_K Y)$, postulating that $\check{f}_R^k(Y,t), \ldots, \check{q}_R^\alpha(Y,t)$ are known functions of $X_k(\bar{Y},t), \ldots, \theta_\alpha(\bar{Y},t)$, $\bar{Y} \in \{Y, Y + \Delta_K Y\}$ (in this case we have to take into account the properties of the material which joint the elements). For a detailed discussion the reader is reffered to [4], where all kinds of iteractions are taken into account.

At the end of the Section we hall give some comments on the problem of relation between the material system M and its discrete model. This model was introduced under two assumptions:

1) material gradients of deformation function and temperature can be treated as constant in each element $\mathscr{P}(Y)$, $Y \in D_R$,

2) interactions among elements are realized only by the continuity of the suitable components of the stress vector and heat flux. The first of these two assumptions leads to the system of reactions r^k , s_R^k , ϑ, w_R in each element, which have to be calculated in each problem we deal with. Such system of reactions does not exist in the material system M; it follows that its magnitude has to be sufficiently small if the discrete system represents a good approximation of the material M(*). Secondly, the interactions among elements in the discrete system defined in this Section lead

(*) Note, that from well known linear approximation of $(1.6)_2$ follows that the relation $T_R^{k\alpha} = 0$ holds in $\mathscr{P}$(Y) only if $\theta_\alpha = 0$. Thus for $\theta_\alpha \neq 0$ and $b_k = 0$ we also obtain the reaction forces $r^k = \mu_R^{-1} T_{R\ ,\alpha}^{k\alpha}$ which are due to the constraints $F_{k\alpha,\beta} = 0$ in $\mathscr{P}$(Y).

to jump discontinuities of the deformation function and temperature because these discontinuities de not exist in M, then their magnitude has to be sufficiently small in any discrete model of M.

3. Constrained thermoelastic continua

From a formal point of view, equations (2.7) – (2.11) can be treated as the finite difference equations of a certain N-th dimensional continuum, provided that D_R is a lattice of points in a region B_R of the space R^N. Denoting by $X = (X^K)$ the points in B_R, and denoting the differentibale functions defined on $B_R \times R$ by the same symbols as the corresponding functions defined on $D_R \times R$, we shall obtain from (2.7) – (2.11) the following system of the field equations:

1. Incompatibility equations for deformations and temperature

$$X_{k,K} - F_{kK} = I_{kK} , \quad \theta_{,K} - \theta_K = i_K , \tag{3.1}$$

where

$$\int_{B_R} I_{kK} \delta T_R^{kK} dv_R = 0 , \int_{B_R} i_K \delta h_R^K dv_R = 0 ; \; dv_R \equiv dX^1, .., dX^N , \quad N < 3, \tag{3.2}$$

have to hold for any continuous fields δT_R^{kK}, δh_R^K consistent with

(3.3) $\alpha^\mu (X, T_R^{kK}) = 0 , \; \mu = 1, ..., m ; \; \beta^\rho (X, h_R^K) = 0 , \; \rho = 1, ..., r,$

α^μ, β^ρ being the known differentiable functions.

2. Equations of motion

$$\begin{aligned} & T_{R\;,K}^{kK} + f_R^k = \rho_R \ddot{X}^k , \\ & \ell^2 I_R^{\alpha\beta} \ddot{F}^k_\beta = \delta_K^\alpha T_R^{kK} - \rho_R \frac{\partial \sigma}{\partial F_{k\alpha}} + S_R^{k\alpha} ; \end{aligned} \tag{3.4}$$

3. Equations of entropy balance

$$h_{R\;,K}^K + \gamma_R + \alpha_R \mathring{\theta} + B_R^{k\beta} \mathring{F}_{k\beta} + \ell a_R^\alpha \dot{\theta}_\alpha = 0 ,$$

$$\delta_K^\alpha h_R^K - \overset{\cdot}{j}_R^{\alpha} + q_R^{\alpha} + \ell a_R^{\alpha} \dot{\theta} + \ell B_R^{\alpha k \beta} \overset{\circ}{F}_{k\beta} + \ell^2 A_R^{\alpha\beta} \dot{\theta}_\beta = 0 . \tag{3.5}$$

The N-th dimensional continuum governed by Eqs. (3.1) – (3.5) will be called the constrained thermoelastic continuum. The stress vector and the flux of heat across the surface in this continuum will be defined in an usual way by virtue of $t_R^k = T_R^{kK} n_{RK}$, $h_R = h_R^K n_{RK}$, respectively, n_R being the unit vector normal to the surface element in $\bar{B}_R$. The lenght parameter ℓ characterizes the weak non-locality of the continuum. Putting $\ell \to 0$, we obtain from $(1.8)_3$ that σ is independent of θ_α (i.e. σ can be interpreted as a strain energy function) and instead of $(3.4)_2$ and $(3.5)_2$ we have

$$\delta_K^\alpha T_R^{kK} = \rho_R \frac{\partial \sigma}{\partial F_{k\alpha}} - S_R^{k\alpha} , \quad \delta_K^\alpha h_R^K = \overset{\cdot}{j}_R^{\alpha} - q_R^{\alpha} , \tag{3.6}$$

while Eq. (3.5) reduce to

$$h_{R,K}^K + \alpha_R \overset{\circ}{\theta} + B_R^{k\alpha} \overset{\circ}{F}_{k\alpha} + \gamma_R = 0 . \tag{3.7}$$

If we postulate also the extra constraints of the form

$$\alpha_\nu (X, X_k, F_{k\alpha}) = 0 , \nu = 1, \dots, n ; \quad \beta_\pi (X, \theta, \theta_\alpha) = 0, \pi = 1, \dots, p, \tag{3.8}$$

then, using an approach analogous to that given in Sec. 2, we obtain

$$f_R^k = \bar{f}_R^k + \lambda^\nu \frac{\partial \alpha_\nu}{\partial X_k} , \quad S_R^{k\alpha} = \bar{S}_R^{k\alpha} + \lambda^\nu \frac{\partial \alpha_\nu}{\partial F_{k\alpha}} ,$$

$$\gamma_R = \bar{\gamma}_R + \mu^\pi \frac{\partial \beta_\pi}{\partial \theta} , \quad q_R^\alpha = \bar{q}_R^\alpha + \mu^\pi \frac{\partial \beta_\pi}{\partial \theta_\alpha} \tag{3.9}$$

where $\lambda^\nu = \lambda^\nu (X, t)$, $\mu^\pi = \mu^\pi (X, t)$ are Lagrange's multipliers for Eqs. (3.8) and $\bar{f}_R^k$, $\bar{\gamma}_R$, $\bar{S}_R^{k\alpha}$, $\bar{q}_R^\alpha$ are known fields characterizing the external agents acting at the system.

Now let us assume that the thermoelastic continuous system defined by Eqs. (3.1) – (3.5) and by suitable boundary conditions: $t_R^k = T_R^{kK} n_{RK}$, $h_R = h_R^K n_{RK}$, is introduced independently of any discrete system. In this case the equations (2.7) – (2.11) can be interpreted as the finite difference equations related to the equations (3.1) – (3.5), in which the lattice parameter ℓ is equal to the lenght parameter characterizing the thermoelastic continuum. If the finite difference equations represent a good approximation of the differential equations of the thermoelastic continuum (the suitable criteria can be found in [1]), and if they represent, at the same time, a good discrete model of the material system M (cf. the final remarks in Sec. 2), then the thermoelastic continuum under consideration can be interpreted as a continuous model of the material system M. It follows that in many cases the continuous models of a material system M can not exist even if there exist the discrete models.

REFERENCES

[1] А.А.САМАРСКИЙ, Введение в теорию разностных схем, изд. "Наука", Москва, 1971.

[2] C. TRUESDELL, W. NOLL, The Non-Linear Field Theories of Mechanics Handbuch der Physik, III/3, Springer-Verlag 1965.

[3] Cz. WOZNIAK, On the problem of constraints for deformations and stresses in continuum mechanics, Bull. Acad. Polon. Sci., ser. sci. techn., 1975.

[4] Cz. WOSNIAK, On the relations between discrete and continuum mechanics, Arch. Mech. Stos., Warszawa, 1976.

LIST OF CONTRIBUTORS

I.N. SNEDDON, Professor of Mathematics at the University of Glasgow, Great Britain.

W. NOWACKI, Vice President of the Polish Academy of Sciences, Warszawa, Poland.

H. PARKUS, Professor of the Institut für Mechanik, Technische Universität, Wien, Austria.

C. WOZNIAK, Professor at the Instytut Mechaniki, Univ. Warszawski, Warszawa, Poland.